T0291911

Saliva Protection and Transmissible Diseases

Saliva Protection and Transmissible Diseases

Jacobo Limeres Posse
University of Santiago de Compostela [USC], Santiago de Compostela, Spain

Pedro Diz Dios
University of Santiago de Compostela [USC], Santiago de Compostela, Spain

Crispian Scully
University College London [UCL], London, United Kingdom

ACADEMIC PRESS

An imprint of Elsevier

Academic Press is an imprint of Elsevier
125 London Wall, London EC2Y 5AS, United Kingdom
525 B Street, Suite 1800, San Diego, CA 92101-4495, United States
50 Hampshire Street, 5th Floor, Cambridge, MA 02139, United States
The Boulevard, Langford Lane, Kidlington, Oxford OX5 1GB, United Kingdom

Copyright © 2017 Elsevier Inc. All rights reserved.

No part of this publication may be reproduced or transmitted in any form or by any means, electronic or
mechanical, including photocopying, recording, or any information storage and retrieval system, without
permission in writing from the publisher. Details on how to seek permission, further information about the
Publisher's permissions policies and our arrangements with organizations such as the Copyright Clearance
Center and the Copyright Licensing Agency, can be found at our website: www.elsevier.com/permissions.

This book and the individual contributions contained in it are protected under copyright by the Publisher
(other than as may be noted herein).

Notices
Knowledge and best practice in this field are constantly changing. As new research and experience broaden our
understanding, changes in research methods, professional practices, or medical treatment may become
necessary.

Practitioners and researchers must always rely on their own experience and knowledge in evaluating and using
any information, methods, compounds, or experiments described herein. In using such information or methods
they should be mindful of their own safety and the safety of others, including parties for whom they have a
professional responsibility.

To the fullest extent of the law, neither the Publisher nor the authors, contributors, or editors, assume any
liability for any injury and/or damage to persons or property as a matter of products liability, negligence or
otherwise, or from any use or operation of any methods, products, instructions, or ideas contained in the
material herein.

British Library Cataloguing-in-Publication Data
A catalogue record for this book is available from the British Library

Library of Congress Cataloging-in-Publication Data
A catalog record for this book is available from the Library of Congress

ISBN: 978-0-12-813681-2

For Information on all Academic Press publications
visit our website at https://www.elsevier.com/books-and-journals

Working together
to grow libraries in
ELSEVIER Book Aid International # developing countries

www.elsevier.com • www.bookaid.org

Publisher: Sara Tenney
Acquisition Editor: Linda Versteeg-Buschman
Editorial Project Manager: Kathy Pallida
Production Project Manager: Anusha Sambamoorthy
Cover Designer: Christian Bilbow

Typeset by MPS Limited, Chennai, India

CONTENTS

AUTHORS' BIOGRAPHIES

Senior Lecturer Jacobo Limeres Posse, DDS, PhD

Jacobo Limeres graduated in Dentistry at the Santiago de Compostela University (USC) and got the PhD degree afterwards. He is the Director of the Stomatology Department of the School of Medicine and Dentistry (USC). Also he is the Coordinator of the Exchange Program for Dental Studies in the USC and the Coordinator of the Program for Dental Care of Severely Handicapped Patients, developed in the Special Needs Unit of the USC in agreement with the Galician Healthcare Service. He is the Co-director of the Postgraduate course, Master in Dentistry for Medically compromised patients at the USC, and currently is the President of the Spanish Society of Dentistry for Handicapped and Special Needs Patients. He is also member of the council of the International Association for Disability and Oral Health (iADH). He has written about 70 peer-reviewed papers and coauthored several books.

Professor Pedro Diz Dios, MD, DDS, PhD, FDSRCS (Hon)

Pedro Diz Dios is qualified in Medicine and is a Specialist in Stomatology at Santiago de Compostela University (Spain). He trained in Oral and Maxillofacial Surgery at the Freiburg University Hospital (Germany), gaining his Doctor of Philosophy PhD (Freiburg University). He is a Professor, Consultant, and Head of the Special Needs Unit, School of Medicine and Dentistry, Santiago de Compostela University. He was a Director of the Stomatology Department and he is a Director of the Master's Program on "Dental management of patients with systemic diseases" at the Santiago de Compostela University. He is a Council member of the International Association for Disability and Oral Health (IADH). He is an Honorary Visiting Professor at the UCL-Eastman Dental Institute (London) and Fellowship (*ad hominem*) of the Royal College of Surgeons of Edimburgh. He is an Editor-in-Chief of *Special Care in Dentistry*, an Associate Editor of *Oral Diseases Journal, Medicina Oral Cirugía Oral y Patología Bucal*, and *Journal of Oral and Experimental Dentistry*, as well as an Editorial Board Member of *Journal of Disability and Oral Health*. He has written about 200 peer-reviewed papers and coauthored several books.

Professor Crispian Scully, CBE, MD, PhD, MDS, MRCS, BSc, FDSRCS, FDSRCPS, FFDRCSI, FDSRCSE, FRCPath, FMedSci, FHEA, FUCL, FSB, DSc, DChD, DMed (HC), Dr.hc (*in memoriam*)

Professor Scully was a Director of the WHO Collaborating Centre for Oral Health-General Health; King James IV Professor at the Royal College of Surgeons of Edinburgh, and Emeritus Professor, University College London (UCL). He was a Specialist in Oral Medicine, in Special Care Dentistry, and in Oral Surgery. He was a Founder member of the European Association for Oral Medicine and a Founder member of the UK Academy of Medical Sciences. He has the distinction of Commander of the British Empire (CBE), and he has been awarded several University Doctorates (Athens, Granada, Helsinki, Pretoria, Santiago de Compostela). Professor Scully was an Editor of *Oral Diseases*, Associate Editor of *Medicina Oral Cirugía Oral Patología Bucal*, and was on the Editorial Boards of *Acta-Oto-Laryngologica and Dental Update*. He founded and was an Editor of *Oral Oncology*. Professor Scully has presented continuing education programs throughout the world and has published over 1000 scholarly works and 45 books (34 authored and 11 edited).

PREFACE

The concept for this book arose and evolved from our long-standing interests in Infective Diseases. These have and will continue to be increasing global issues, from the plagues of former times to recent outbreaks from the emergence of formerly isolated diseases such as HIV to Ebola or Zika. Most transmission is via people, partners, practices, or pets. Even before birth, the fetus can be infected by a range of agents from syphilis to TORCH organisms (Toxoplasmosis, Rubella, Cytomegalovirus, Herpes) and Zika—with sometimes devastating consequences.

This book should be of interest to Health and Social care personnel as well as the public, reviewing infections affecting or arising from the mouth microorganisms potentially transmissible by saliva, and considering the evidence on diseases that may be transmitted by kissing. Space precludes the involvement with animals and the potential future use of salivary markers since there are often other reliable tests in use so most are experimental only.

Jacobo Limeres Posse[1], Pedro Diz Dios[1] and Crispian Scully[2]
[1]Santiago de Compostela University, Santiago de Compostela, Spain
[2]University College London [UCL], London, United Kingdom

Saliva is a complex body fluid obviously essential to health, especially mastication, swallowing, and speech. The instinctive licking of wounds in humans and animals suggests some underlying innate belief in healing properties. Mechanical cleansing and innate and acquired immunity underlie protection. Hyposalivation leads to dysfunction and also to liability to infections, but there can be complex interactions between microorganisms and protective factors.

Kissing is a feature of daily life. In many cultures, human kissing is a feature of greetings, particularly among family members. There are also situations in which kissing can have sexual connotations. Kissing may sometimes imply an exchange of saliva, making it an activity that can potentially produce person-to-person transmission of microorganisms. Exchange of saliva may also be seen in premastication of foods for newborns and infants. Few people show concern about the possibility of infection via saliva, or perhaps they chose to disregard it.

Even in instances where ductal saliva might contain no microbial agents, whole saliva may contain agents from the exogenous introduction of microorganism via fingers, utensils and other objects, and from endogenous sites—most notably blood or gingival crevicular fluid, and the respiratory tract.

In addition to the dentally important cariogenic and periodontopathogenic flora, microorganisms responsible for a range of diseases involving bacteria, viruses, fungi, and other organisms such as tuberculosis, syphilis, meningococcal meningitis, *herpesviruses*, and *papillomaviruses* can be conveyed by saliva and by kissing. Many of these infections are seen especially in immunocompromised people and can be life-changing or lethal. The emergence of dengue, chikungunya, Ebola, and other viruses such as Zika highlights this area for both health professionals and public globally.

Most evidence of microbial transmission comes from epidemiological studies but it may be difficult to differentiate saliva transmission from transmission by fomites or body fluids introduced into the mouth, respiratory droplets, aerosols, or other routes. Epidemiological data and the presence of saliva in nucleic acid or antigens of microorganism have to be relied upon as circumstantial evidence. Reliable studies designed specifically to clarify which diseases can be transmitted in humans by human saliva are needed in order to develop appropriate strategies to control person-to-person spread of infection by this route.

CHAPTER 1

Infection Transmission by Saliva and the Paradoxical Protective Role of Saliva

1.1 SALIVA COMPOSITION AND SECRETION

Saliva is produced by both major (parotid and submandibular and sublingual) and minor (located in the mouth) glands, with different constituents and properties between the two groups. In the mouth saliva is a colorless, odorless, tasteless, watery liquid containing 99% water and 1% organic and inorganic substances and dissolved gases, mainly oxygen and carbon dioxide. Salivary constituents can be grouped into proteins (e.g., amylase and lysozyme), organic molecules (e.g., urea, lipids, and glucose mainly), and electrolytes (e.g., sodium, calcium, chlorine, and phosphates).[1] Cellular elements such as epithelial cells, leucocytes and various hormones, and vitamins have also been detected. The composition of saliva is modified, depending on factors such as secreted amount, circadian rhythm, duration and nature of stimuli, diet, and medication intake, among others.

Despite this heterogeneous composition, from the functional point of view saliva has to be considered as a unique biological fluid, and not as the sum of its biochemical components.[2,3]

Salivary secretion and maintenance of a film of saliva on oral surfaces is dependent upon nerve-mediated, reflex salivary gland secretion mainly stimulated by taste. The afferent arm is mainly activated by stimulation of chemoreceptors (located in the taste buds) and mechanoreceptors (located in the periodontal ligament).[4] Olfaction, mental processes, and stretch of the stomach are weak stimuli. Impulses affecting secretion depending on the emotional state are carried by afferent cranial nerves V, VII, IX, and X to the CNS salivary nuclei (salivation center) in the medulla oblongata. The efferent part of the reflex is mainly parasympathetic. The cranial nerve VII provides control of the submandibular, sublingual, and minor glands, whereas the cranial nerve IX controls the parotid glands. The flow of saliva is enhanced

Saliva Protection and Transmissible Diseases. DOI: http://dx.doi.org/10.1016/B978-0-12-813681-2.00001-9
© 2017 Elsevier Inc. All rights reserved.

by sympathetic innervation, which promotes contraction of muscle fibers around the salivary ducts.[5] Autonomic nerves also have an important role in both gland development and function.[6] A dry mouth is a common experience where there is fear.

Saliva may be secreted in the absence of exogenous stimuli, then referred to as the resting or unstimulated salivary flow. In the resting state 70% of saliva is secreted by the submandibular and sublingual glands. When stimulated, the parotids provide most of the saliva and flow can increase by up to fivefold. On average, in healthy nonmedicated adults, the unstimulated and chewing-stimulated salivary flow rates are about 0.3 and 1.5 mL/min respectively,[1] but the range is wide and the limits of normality in all age groups and both genders are considerable. The normal daily production of saliva varies from 700 mL to 1.5 L. A decrease in the daily production of saliva below 500 mL/ day is termed hyposecretion or hyposialia.[7] Sialorrhea, hypersialia, hypersalivation, and ptyalism are terms used to describe salivary flow above the limits of the normal.[8]

Saliva plays a central role in oral health monitoring, regulating and maintaining the integrity of the oral hard and soft tissues.[1] It lubricates and cleans the oral cavity, possesses antibacterial, antiviral, and antifungal properties, buffers the pH, helps in chewing, speech, swallowing, and digestion, promotes taste, and contributes to the maintenance and remineralization of teeth.[9] Moreover, it may be useful in the diagnosis of various diseases.[10] The characterization by proteomic approaches— of more than 1000 salivary proteins and peptides—has allowed the identification of new salivary markers in oncology, salivary gland dysfunction, Sjögren's syndrome, systemic sclerosis, psychiatric and neurological diseases, and dental and periodontal pathology.[11-13]

1.2 INFECTION TRANSMISSION BY SALIVA AND KISSING

The infectivity of microorganisms can depend on the infective load, virulence, with some, such as the notorious *norovirus*, being extremely contagious and able to survive weeks on surfaces and fomites.[14] The detection and continuous shedding of infectious agents in saliva does not necessarily mean transmission by this route. Factors including the microorganism load, the existence of specific receptors on oral epithelial cells, and host defenses may play an important defensive role.[15] Moreover, blood contamination in saliva—often invisible to the eye—is not uncommon mainly among active smokers[16] and individuals with poor oral health status, those with gingivitis or periodontitis,[17] and those with certain infectious diseases including human immunodeficiency virus (HIV) infection.[18]

Saliva contact can cause overt concern when using utensils such as cutlery or oral health devices, or if kissing a person with an infectious disease. However, the apparent absence of obvious disease does not guarantee the absence of infection or infective agents in saliva (or other body fluids): many diseases (especially viral) can be incubating or be subclinical (causing no or nonspecific symptoms or signs). Intimate mucosal contacts, particularly where there are epithelial breaches or substances that may impede salivary defenses (e.g., other body fluids), predispose to infection transmission.

Kissing is not exclusive to humans or primates, though it may have different connotations in different species.[19] Theories to explain kissing behavior consider it to have an origin in social and sexual interactions, premastication of foods for newborns or even the intentional transfer of microorganisms to promote immunity.[20] Kissing is seen in most human cultures,[20] and often is part of daily behavior, playing important roles in building and maintaining interpersonal relationships[21,22] and in partner selection.[23] There are, however, huge intercultural differences related to kissing; this being considered an acceptable behavior in some cultures but totally offensive in others. For example, social kissing is an accepted form of salutation in the Mediterranean and Latin cultures, in Muslim-majority societies governed by religious law there are strict taboos about whom one can kiss, or people from some areas in Sudan refuse to kiss because they fear having their soul stolen through kissing.[24] In general, kissing is considered by many of the public to have few or no serious health implications.

Different types of kissing are evident and the type of kissing may well be relevant with respect to the transmission of microorganisms, as it not only determines the capacity of the kiss to spread infectious diseases,[25] but it can also have a bearing on the chemoprophylaxis strategy to be used in "kissing contacts" in certain situations (e.g., during an outbreak of meningococcal disease).[26]

"Air kissing" is a cheek-to-cheek approximation; "osculum" is when the lips make contact with the body, usually the cheeks; "basium kiss" consists of mutual approximation of the lips with the mouth closed, exercising light pressure; and finally, there is the "saviolum kiss" in which, in addition to lip contact, the tongue is inserted into the opposite person's mouth ("French kissing," "passionate kissing," deep kissing," "active kissing," or "intimate kissing").[27] Finally "kiss of life" refers to direct, intense, and recurrent lip contact during mouth-to-mouth resuscitation—the therapy of choice for cardiorespiratory arrest in the community.

Couples may exchange an average of 5 mL of saliva during active kissing,[28] making this an activity that could favor the transmission of infectious diseases. Evidence for person-to-person transmission by kissing is limited to a few microorganisms and even this evidence can be often based on only weak scientific evidence. Published studies are heterogeneous, from isolated case reports of kissing as a "possible" cause of transmission of diseases (e.g., HIV infection)[29] to studies analyzing the inhibitory activity of the saliva on specific microorganisms (e.g., *herpes simplex*).[30] Few studies have been designed specifically to demonstrate the degree of infectivity if any of kissing, but one study showed it does not efficiently spread common cold infection by *Rhinoviruses*.[31]

Studies on the risks from mouth-to-mouth ventilation without barrier devices[32] demonstrated isolated cases of transmission of tuberculosis,[33] *herpes simplex* infection,[34] shigellosis,[35] salmonellosis,[36] and meningococcal infection.[37]

Despite this, evidence for infection transmission by kissing is not strong, so this does not justify philemaphobia (morbid fear of kissing). Paradoxically, it has even been suggested that kissing could be an evolutionary adaptation to protect against some neonatal infections (e.g., *cytomegalovirus*).[38] In reality, saliva may also have a protective role, and many animals, even humans, instinctively lick wounds—an act that may be defensive—possibly via histatins mainly.[39,40] Saliva also contains an array of factors which facilitate protection (Table 1.1).

1.3 THE PROTECTIVE ROLE OF SALIVA

Adequate salivary flow has a cleansing action and saliva also contains potentially protective constituents (Table 1.1).[40–42] Antimicrobial

Table 1.1 Antimicrobial Factors in Saliva

Factor	Antibacterial	Antiviral	Main details
Agglutinins	+	+	gp340, DMBT1 (deleted in malignant brain tumors 1)
Antibodies	+	+	sIgA (secretory Immunoglobulin A)
Calgranulin or calprotectin	+	−	Calgranulin A/B is antimicrobial by binding calcium and other metals
Cathelcidin	+	+	Cathelcidin is cleaved into the antimicrobial peptide LL-37 by both kallikrein 5 and kallikrein 7 serine proteases
Cystatin	+	+	A cysteine proteinase inhibitor which can be antiviral
Defensins	+	+	HNPs (Human Neutrophil Peptides) α,β
Histatins	+	−	A family of histidine-rich antimicrobial proteins, especially antifungal
Lactoferrin	+	+	An iron-binding glycoprotein in saliva and various other secretory fluids
Lysozyme (muraminidase)	+	−	Damages bacterial cell walls
Mucins	+	+	Glycoconjugates (glycosylated proteins) produced by epithelia. Membrane-associated mucins may also act as cell surface receptors for pathogens
Peroxidase	+	+	Produced mainly by parotid gland
Secretory leukocyte protease inhibitor	+	+	SLPI is found in saliva and many other secretions, protects epithelial tissues from serine proteases, and is antimicrobial

proteins can arise from epithelial cells, innate immune, and other cells and can modulate the microbial flora in the mouth. For example, viruses such as *noroviruses* are affected by host genetic factors [43] including histoblood group antigens (HBGAs) (i.e., the ABO blood group, the Lewis phenotype, and the secretor status).

Salivary proteins which can be protective at least against certain agents, include scavenger receptor cysteine-rich glycoprotein 340 (salivary gp-340), mucins, histatins, and human neutrophil defensins. The protein gp340—formerly salivary agglutinin—aggregates a variety of bacteria and can function as a specific inhibitor of HIV-1 and influenza A.[41] Salivary mucins MUC5B and MUC7 reduce the attachment and biofilm formation of *Streptococcus mutans* by keeping bacteria in the planktonic state.[44] Several studies have shown that salivary mucins induce phenotypic changes in *Candida albicans* at the level of mRNA transcription, which downregulate genes necessary for hyphal development and some virulence factors.[45] Saliva also contains an array of other protective proteins including tissue factor, growth factors—

especially Secretory Leukocyte Protease Inhibitor (SLPI) and Epidermal Growth Factor—which may in addition, facilitate wound healing.[39]

It has been reported that saliva inhibits oral transmission of HIV through kissing, dental treatment, biting, and aerosolization; both crude saliva and mucins MUC5B and MUC7 inhibit HIV-1 activity, probably because they trap or aggregate the virus and prevent its entry into host cells.[46] SLPI is also important. *Hantaviruses* are also sensitive to the antiviral actions of mucins,[47] and sialic acid type molecules have high activity against *human influenza viruses*.[48]

Histatins provide the first line of defense against *C. albicans*[49] and other fungi.[50] Cystatin may inhibit *coronaviruses*.[51] Moreover, saliva mediates antibody-dependent cell-mediated cytotoxicity as in HIV-1-infected individuals[52] and can regulate specific humoral defense mechanisms against microorganisms including *Cryptococcus neoformans*[53] or *Paracoccidioides*.[54]

Oral carriage of microorganisms and infections are more likely where there is hyposalivation and/or immunoincompetence—and so infections may be more prevalent in neonates who lack acquired immunity, or where immunity wanes such as in older or patients with immunocompromising conditions (e.g., malignant disease and its treatment, HIV/AIDS, or corticosteroid therapy)—particularly where the load of infecting agents is high or the microbe is virulent.

- *Specific saliva protection against oral bacteria*
  Saliva has a mechanical flushing action and there are innate immune defenses and complex interactions with microorganisms.[55,56] For example, the gene DMBT1 (Deleted in Malignant Brain Tumor 1) encodes antimicrobial proteins involved in mucosal innate immunity, and salivary DMBT1 glycoprotein (gp340) and salivary agglutinin (DMBT1(SAG)) glycoproteins which are identical, agglutinate *S. mutans* and some other Gram-positive bacteria, as well as several Gram-negative bacteria.[57] Some of the salivary components can change with disease; e.g., higher interleukin (IL)-6/IL-1β, secretory IgA, and lower lysozyme, and histatins 1 and 5 have been found in hepatic cirrhosis.[58]
  The innate and acquired immune defenses in saliva persist even after removal of lymphoid tissue in tonsillectomy: serum-derived

antimicrobial proteins (myeloperoxidase, lactoferrin, IgG) remain in high concentrations in whole saliva with no effect on the numbers of oral cariogenic *S. mutans* or on the total aerobic flora.[59]

- *Specific saliva protection against bacteria such as Pseudomonas aeruginosa*

Pseudomonas aeruginosa often colonizes the airways in cystic fibrosis. *P. aeruginosa* binds to oral and bronchial epithelial cells,[60,61] by pili and fimbriae which promote adherence to glycosphingolipid adhesins asialo-GM1 on surfaces of host epithelia and phagocytes such as polymorphonuclear leukocytes.[62–64] Failure to isolate pathogenic organisms consistently from the upper airways in all patients with positive sputum argues against a local epithelial factor predisposing to bacterial colonization[65] and also suggests that defensive processes are in play. *P. aeruginosa* are aggregated by saliva.

The sero-mucous products of the submandibular gland have a greater role than the serous secretions of the parotids and are possibly responsible for the differences in oral colonization by *P. aeruginosa* in different subjects.[66] The low-molecular-weight mucin (MG2) of human submandibular–sublingual saliva, and neutral cystatin, bind to pili.[67] *P. aeruginosa* interactions with *S. aureus* may be predicated on the formation of MG2–secretory IgA antibody complex, which may facilitate clearance from the oral cavity.[68]

Interbacterial adherence between strains of *P. aeruginosa* with oral *Actinomyces viscosus* indigenous to the human mouth and with strains of *Streptococcus pyogenes*, and *Streptococcus agalactiae*, appear to involve galactosyl-binding adhesins.[69] Most oral *viridans streptococci* have potentially bacteriocin-like activity against *P. aeruginosa*.[70]

- *Specific saliva protection against viruses such as HIV*

Saliva may also be inhibitory to HIV. Though complete inactivation may require 30 minutes of exposure, saliva may inhibit HIV-1.[71–76] A main protective mechanism of saliva may be the inactivation of HIV-transmitting leukocytes by the hypotonicity of saliva[77] and the oral transmission of HIV by seminal and other fluids introduced into the mouth may be due to their isotonicity overcoming the inactivation of HIV by isotonic saliva.[78] HIV transmission across mucosae involves complex mechanisms and the oral mucosal epithelia mucosa is less permissive for HIV replication than other sites (e.g., vagina/cervix and anal/rectal).[79] Innate immunity plays a role in protection.[80] Viral reception appears to involve both CD4 (Cluster of Differentiation 4) and a co-receptor—particularly CCR5

(Chemokine Receptor type 5).[81] The MHC appears to have a role in HIV-1 control, particularly the HLA Complex P5 (HCP5) and Human Leukocyte Antigen-C (HLA-C) and this may explain the occurrence of "Elite Supressor" patients.[82]

The scavenger receptor protein gp340—encoded by the DMBT1 gene—interacts with surfactant proteins (SP-D), and both gp340 and SP-D can individually and together interact and agglutinate some viruses and DMBT1(gp340) binds to a variety of other host proteins, including serum and secretory IgA, Clq, lactoferrin, MUC5B, and Trefoil Factor 2 (TFF2), all molecules involved in innate immunity and/or wound healing.[57] The protein gp340 appears to facilitate HIV transmission across genital but not oral mucosa.[83–86] Acquired immunity might confer some protection in re-exposures: immunization with an HIV peptide may produce HIV-inhibitory antibodies in saliva.[87]

Various glycoproteins may also be inhibitory to HIV. Crude saliva and salivary mucins MUC5B and MUC7 (both from HIV-positive people and uninfected controls) can inhibit HIV-1.[46,88] Other glycoproteins may also be implicated.[89,90]

SLPI may have an important HIV-inhibitory role,[91,92] as might human β-defensins (hBDs) from the epithelium.[93,94] Saliva may also mediate antibody-dependent cytotoxicity against HIV.[52]

- *Specific saliva protection against influenza A virus*

Other viruses as. e.g., H5N1 *influenza virus* are particularly susceptible to human saliva, which may play a role in its infectivity and transmissibility.[48]

Many salivary antibacterial proteins have antiviral activity, typically against specific pathogens.[41] Antiviral activities of saliva against influenza A virus (IAV) and HIV differ both in terms of specific glandular secretions and the inhibitory proteins. Whole saliva or parotid or submandibular/sublingual secretions from healthy donors inhibits IAV, whereas only submandibular/sublingual secretions are inhibitory to HIV. Among salivary proteins, scavenger receptor cysteine-rich glycoprotein 340 (gp340), MUC5B, histatins, and human neutrophil defensins at concentrations present in whole saliva inhibit IAV, while acidic proline-rich proteins and amylase have no activity nor do several less abundant salivary proteins (e.g., thrombospondin or serum SLPI).[95]

gp340 interacts with surfactant proteins A and D (SP-D) and can interact and agglutinate IVA virus and also binds to proteins involved in innate immunity and/or wound healing, including serum and secretory IgA, C1q, lactoferrin, MUC5B and TFF2.[57] Salivary gp340 can antagonize SP-D antiviral activities—which may be relevant to the effects of aspiration of oral contents on SP-D-mediated lung functions.[96]

Other components responsible for antiviral activity on *influenza virus*, in particular *swine origin influenza A virus* (S-OIV), include an α-2-macroglobulin (A2M) and an A2M-like protein.[97]

Salivary glycoproteins which have significant roles against IVA also include lectins (e.g., MAL-II and SNA).[98]

MUC5B inhibits IAV by presenting a sialic acid ligand for the viral hemagglutinin.[95] Other sialic acid–containing molecules may be effective against *human influenza viruses* more so than against H5N1.[48]

- *Specific saliva protection against fungi such as Candida* spp.

Both innate immunity and cell-mediated immune response are involved in defenses against fungal infections. Saliva has a mechanical defense action and components including secretory immunoglobulin A, lactoferrin, and polymorphonuclear leukocyte (PMNL) superoxide are protective.[99] A low, stimulated salivary flow rate—not a low, unstimulated flow rate—is associated with *Candida* spp. carriage.[100]

Salivary components mediate microbial attachment to oral surfaces and interact with planktonic microbial surfaces to facilitate agglutination often mediated by lectin-like proteins that bind to glycan motifs on salivary glycoproteins and help eliminate pathogens. Antimicrobial peptides in saliva appear to play a crucial role in the regulation of oral *Candida* growth. Oral candidiasis may be associated with salivary gland hypofunction and decreases of salivary lactoferrin, secretory immunoglobulin A, β-defensin 1, and β-defensin 2 antibacterial proteins.[101]

Histatins are basic histidine-rich cationic proteins present in saliva that provide the first line of defense against oral candidiasis—an important antimicrobial is histatin 5 (Hst 5),[102,103] which shows potent and selective antifungal activity and with the carrier molecule spermidine which, by binding to fungal cell wall proteins (Ssa1/2) and glycans, significantly enhances *C. albicans* killing.[49] Histatins effectively kill *C. albicans, C. glabrata,* and *C. Krusei,* and histatin 3 acts against *C. dubliniensis.*[104]

Other antimicrobial proteins include calprotectin,[105] cystatin SA1,[106] and β-defensin 2,[107] deficiencies of which predispose to chronic candidiasis. Salivary lysozyme can also be protective.[108]

The development of candidiasis in HIV-infected patients could be a consequence of inefficient lysozyme and lactoferrin concentrations and of decreased cooperation between innate and adaptive immune systems.[109] The vast majority of *Candida* isolates appear to succumb to nonspecific host immune mediators[110] but innate immunity alone is unable to stop yeast expansion in HIV-infected patients.[111]

C. albicans-secreted aspartyl proteinase (SAP1-SAP8) and phospholipase B (PLB1 and PLB2) genes are expressed during both infection and carriage of *Candida* spp. The differential expression of these hydrolytic enzyme genes correlates the expression of specific *Candida* spp. virulence genes with active candidiasis and anatomical location.[112] Salivary anti-somatic, anti-SAP2, and anti-SAP6 antibodies are not efficient in limiting candidal infection[113] and although HIV-infected patients have a high mucosal response against *C. albicans* virulence antigens, such as somatic antigen, Sap1, and Sap6,[114] this is not totally protective.

Defensins such as α-defensin (Human Neutrophil Peptides, HNPs) and β-defensin-2 (hBD-2) peptides can have antifungal and cytotoxic activities.[115] Defensins that exhibit antibacterial, antifungal, and antiviral properties are a component of the innate immune response. β-defensins (hBD-1) are cationic antimicrobial peptides encoded by the DEFB1 gene expressed in oral epithelia that may have a major role in mediating and/or contributing to susceptibility to candidiasis.[116]

Nitric oxide (NO) is involved in host resistance to infection with *C. albicans* at least in animal models. IL-4 is associated with resistance to oral candidiasis and suggests that NO is involved in controlling colonization of the oral mucosal surface with *C. albicans*.[117]

Oral epithelial cells may play a role in innate resistance against candidiasis.[118] Host defenses against *C. albicans* include epithelial cell defenses and innate and specific immune mechanisms. Cell-mediated immunity by Th1-type CD4 + T-cells is important for protection against mucosal infections, and PMNLs are important for protection against systemic infections.[119] When CD8(+) T-cell migration is inhibited by reduced tissue E-cadherin, there is susceptibility to infection which supports a role for CD8(+) T cells in host defense against oropharyngeal candidiasis.[120]

Fungal pattern recognition receptors such as C-type lectin receptors trigger protective T-helper (Th)17 responses in the oral mucosa. The Th17/IL-17 axis is vital for immunity to fungi, especially *C. albicans.* The inflammatory cytokine IL-17 induces tumor necrosis factor (TNF)-α, and interleukins IL-1β and IL-6.[121] A systemic immune response involving T-helper 1 (Th1) cells with the production of TNF-α and IFN-γ is seen in patients with oral candidiasis.[122] Th17 cells may act through IL-17, to confer defenses via neutrophils and antimicrobial factors.[123] Oral epithelial cells also are involved in local host defenses against *C. albicans* infections via IFN-γ induced IL-18.[124]

Biofilms, some 15% of which may be due to dual *Candida* spp., contribute to the pathogenesis of oral candidiasis,[125] biofilm formation of *C. albicans* appearing to be modulated by salivary and dietary factors.[126]

C. albicans hyphal wall protein 1 (Hwp1) mRNA is present in candidiasis regardless of symptoms, implicating hyphal and possibly pseudohyphal forms in mucosal carriage as well as disease.[127] Overall, Hwp1 and hyphal growth forms appear to be important factors in both benign and invasive interactions of *C. albicans* with human hosts.

1.4 PREVENTION OF TRANSMISSION OF MICROORGANISMS BY SALIVA

Transmission of infection by saliva may be prevented or minimized by avoidance of exposure, by good oral hygiene (plaque removal), and by the use of the various substances such as some mouthwashes, and probiotics that may inhibit salivary microorganisms.[128,129]

1.5 CLOSING REMARKS AND PERSPECTIVES

Bacterial pathogens have been identified in salivary samples by specific antibody reactivity, antigen detection, or via PCR, including *Escherichia coli, Mycobacterium tuberculosis, Treponema pallidum,* and a wide range of *Streptococcus* spp. More than 20 viruses have also been detected; these include a number of *Herpes viruses, Hepatitis viruses, Human Immunodeficiency Viruses, Papillomavirus, Influenza virus, or Poliovirus.* Nonviral and nonbacterial infectious agents including fungi and protozoa are also detectable, usually by antibodies to these infectious agents. Recognition of the components of the oral microbiota can help in the prediction of the onset, progression, and prognosis of oral and systemic diseases. Tests for these pathogens are currently under development. Omics methods, such as 16S rRNA sequencing, metagenomics, and metabolomics, can play an essential role to explore microbial community and its metabolite production, without the biases of microbial culture. Saliva contains many antibacterial, antiviral, and antifungal agents which modulate the oral microbial flora. Consequently, detection and shedding of infectious agents in saliva does not necessarily mean transmission by this route. Anyway, the presence of these pathogens in saliva is particularly important in immunosuppressed patients in whom infections can result fatal. Moreover, the defensive ability of saliva against emerging infectious diseases caused by new or previously unrecognized microorganisms remains unknown.

REFERENCES

1. Sreebny LM. Saliva in health and disease: an appraisal and update. *Int Dent J* 2000;**50**(3):140−61.

2. Roth GI, Calmes RB. Salivary glands and saliva. In: Roth GI, Calmes RB, editors. *Oral biology.* St. Louis, MO: Mosby; 1981. p. 196−236.

3. Edgar WM. Saliva: its secretion, composition and functions. *Br Dent J* 1992;**172**(8):305−12.

4. Pedersen AM, Bardow A, Jensen SB, Nauntofte B. Saliva and gastrointestinal functions of taste, mastication, swallowing and digestion. *Oral Dis* 2002;**8**(3):117−29.

5. Hockstein NG, Samadi DS, Gendron K, Handler SD. Sialorrhea: a management challenge. *Am Fam Phys* 2004;**69**(11):2628−34.

6. Proctor GB, Carpenter GH. Salivary secretion: mechanism and neural regulation. *Monogr Oral Sci* 2014;**24**:14−29.

7. Jenkins G. Saliva. In: Jenkins GN, editor. *The physiology and biochemistry of the mouth.* 4th ed. Oxford: Blackwell Scientific Publications; 1978. p. 284−359.

8. Scully C, Limeres J, Gleeson M, Tomas I, Diz P. Drooling. *J Oral Pathol Med* 2009;**38**(4):321−7.

9. Dawes C. Physiological factors affecting salivary flow rate, oral sugar clearance, and the sensation of dry mouth in man. *J Dent Res* 1987;**66**:648−53 Spec No

10. Malamud D. Saliva as a diagnostic fluid. *BMJ* 1992;**305**(6847):207−8.

11. Castagnola M, Picciotti PM, Messana I, et al. Potential applications of human saliva as diagnostic fluid. *Acta Otorhinolaryngol Ital* 2011;**31**(6):347−57.

12. Nunes LA, Mussavira S, Bindhu OS. Clinical and diagnostic utility of saliva as a noninvasive diagnostic fluid: a systematic review. *Biochem Med (Zagreb)* 2015;**25**(2):177−92.

13. Podzimek S, Vondrackova L, Duskova J, Janatova T, Broukal Z. Salivary markers for periodontal and general diseases. *Dis Markers* 2016;**2016** 9179632

14. Goodgame R. Norovirus gastroenteritis. *Curr Gastroenterol Rep* 2006;**8**(5):401–8.

15. Ferreiro MC, Dios PD, Scully C. Transmission of hepatitis C virus by saliva? *Oral Dis* 2005;**11**(4):230–5.

16. Kim YJ, Kim YK, Kho HS. Effects of smoking on trace metal levels in saliva. *Oral Dis* 2010;**16**(8):823–30.

17. Kamodyova N, Banasova L, Jansakova K, et al. Blood contamination in saliva: impact on the measurement of salivary oxidative stress markers. *Dis Markers* 2015;**2015** 479251

18. Piazza M, Chirianni A, Picciotto L, Cataldo PT, Borgia G, Orlando R. Blood in saliva of patients with acquired immunodeficiency syndrome: possible implication in sexual transmission of the disease. *J Med Virol* 1994;**42**(1):38–41.

19. de Waal FB. Primates—a natural heritage of conflict resolution. *Science* 2000;**289** (5479):586–90.

20. Kirshenbaum S. *The science of kissing: what our lips are telling us.* Hachette, UK: Grand Central Publishing; 2011.

21. Hughes SM, Harrison MA, Gallup GG. Sex differences in romantic kissing among college students: an evolutionary perspective. *Evol Psychol* 2007;**5**(3):612–31.

22. Wlodarski R, Dunbar RI. What's in a kiss? The effect of romantic kissing on mating desirability. *Evol Psychol* 2014;**12**(1):178–99.

23. Wlodarski R, Dunbar RI. Examining the possible functions of kissing in romantic relationships. *Arch Sex Behav* 2013;**42**(8):1415–23.

24. Poyatos F. Kinesics: gesturers, manners and postures. In: Poyatos F, editor. *Nonverbal communication across disciplines: Volume 2: Paralanguage, kinesics, silence, personal and environmental interaction.* Amsterdam: John Benjamins Publishing; 2002. p. 216–30.

25. Willmott FE. Transfer of gonococcal pharyngitis by kissing? *Br J Vener Dis* 1974;**50** (4):317–18.

26. Hayward A. Carriage of meningococci in contacts of patients with meningococcal disease. "Kissing contacts" need to be defined. *BMJ* 1999;**318**(7184):665.

27. Touyz L. Lips, kissing and oral implications. *J Aesthet Dent* 2009;**3**(5):29–34.

28. Woolley R. The biologic possibility of HIV transmission during passionate kissing. *JAMA* 1989;**262**(16):2230.

29. Anonymous. Kissing reported as possible cause of HIV transmission. *J Can Dent Assoc* 1997;**63**(8):603.

30. Mikola H, Waris M, Tenovuo J. Inhibition of herpes simplex virus type 1, respiratory syncytial virus and echovirus type 11 by peroxidase-generated hypothiocyanite. *Antiviral Res* 1995;**26**(2):161–71.

31. D'Alessio DJ, Meschievitz CK, Peterson JA, Dick CR, Dick EC. Short-duration exposure and the transmission of rhinoviral colds. *J Infect Dis* 1984;**150**(2):189–94.

32. Giammaria M, Frittelli W, Belli R, et al. Does reluctance to perform mouth-to-mouth ventilation exist among emergency healthcare providers as first responders?. *Ital Heart J Suppl* 2005;**6**(2):90–104.

33. Heilman KM, Muschenheim C. Primary cutaneous tuberculosis resulting from mouth-to-mouth respiration. *N Engl J Med* 1965;**273**(19):1035–6.

34. Hendricks AA, Shapiro EP. Primary herpes simplex infection following mouth-to-mouth resuscitation. *JAMA* 1980;**243**(3):257–8.

35. Todd MA, Bell JS. Shigellosis from cardiopulmonary resuscitation. *JAMA* 1980;**243**(4):331.

36. Ahmad F, Senadhira DC, Charters J, Acquilla S. Transmission of *Salmonella* via mouth-to-mouth resuscitation. *Lancet* 1990;**335**(8692):787−8.

37. Feldman HA. Some recollections of the meningococcal diseases. The first Harry F. Dowling lecture. *JAMA* 1972;**220**(8):1107−12.

38. Hendrie CA, Brewer G. Kissing as an evolutionary adaptation to protect against human cytomegalovirus-like teratogenesis. *Med Hypotheses* 2010;**74**(2):222−4.

39. Oudhoff MJ, Bolscher JG, Nazmi K, et al. Histatins are the major wound-closure stimulating factors in human saliva as identified in a cell culture assay. *FASEB J* 2008;**22**(11):3805−12.

40. Brand HS, Ligtenberg AJ, Veerman EC. Saliva and wound healing. *Monogr Oral Sci* 2014;**24**:52−60.

41. Malamud D, Abrams WR, Barber CA, Weissman D, Rehtanz M, Golub E. Antiviral activities in human saliva. *Adv Dent Res* 2011;**23**(1):34−7.

42. Dawes C, Pedersen AM, Villa A, et al. The functions of human saliva: a review sponsored by the world workshop on oral medicine VI. *Arch Oral Biol* 2015;**60**(6):863−74.

43. Le Pendu J, Nystrom K, Ruvoen-Clouet N. Host-pathogen co-evolution and glycan interactions. *Curr Opin Virol* 2014;**7**:88−94.

44. Baughan LW, Robertello FJ, Sarrett DC, Denny PA, Denny PC. Salivary mucin as related to oral *Streptococcus mutans* in elderly people. *Oral Microbiol Immunol* 2000;**15**(1):10−14.

45. Kavanaugh NL, Zhang AQ, Nobile CJ, Johnson AD, Ribbeck K. Mucins suppress virulence traits of *Candida albicans*. *MBio* 2014;**5**(6):e01911−14.

46. Habte HH, Mall AS, de Beer C, Lotz ZE, Kahn D. The role of crude human saliva and purified salivary MUC5B and MUC7 mucins in the inhibition of human immunodeficiency virus type 1 in an inhibition assay. *Virol J* 2006;**3**:99.

47. Hardestam J, Petterson L, Ahlm C, Evander M, Lundkvist A, Klingstrom J. Antiviral effect of human saliva against hantavirus. *J Med Virol* 2008;**80**(12):2122−6.

48. Limsuwat N, Suptawiwat O, Boonarkart C, Puthavathana P, Wiriyarat W, Auewarakul P. Sialic acid content in human saliva and anti-influenza activity against human and avian influenza viruses. *Arch Virol* 2016;**161**(3):649−56.

49. Puri S, Edgerton M. How does it kill? Understanding the candidacidal mechanism of salivary histatin 5. *Eukaryot Cell* 2014;**13**(8):958−64.

50. Hanasab H, Jammal D, Oppenheim FG, Helmerhorst EJ. The antifungal activity of human parotid secretion is species-specific. *Med Mycol* 2011;**49**(2):218−21.

51. Collins AR, Grubb A. Cystatin D, a natural salivary cysteine protease inhibitor, inhibits coronavirus replication at its physiologic concentration. *Oral Microbiol Immunol* 1998;**13**(1):59−61.

52. Kim HN, Meier A, Huang ML, et al. Oral herpes simplex virus type 2 reactivation in HIV-positive and -negative men. *J Infect Dis* 2006;**194**(4):420−7.

53. Igel HJ, Bolande RP. Humoral defense mechanisms in cryptococcosis: substances in normal human serum, saliva, and cerebrospinal fluid affecting the growth of *Cryptococcus neoformans*. *J Infect Dis* 1966;**116**(1):75−83.

54. Miura CS, Estevao D, Lopes JD, Itano EN. Levels of specific antigen (gp43), specific antibodies, and antigen-antibody complexes in saliva and serum of paracoccidioidomycosis patients. *Med Mycol* 2001;**39**(5):423−8.

55. Scannapieco FA. Saliva-bacterium interactions in oral microbial ecology. *Crit Rev Oral Biol Med* 1994;**5**(3-4):203−48.

56. Marsh PD, Do T, Beighton D, Devine DA. Influence of saliva on the oral microbiota. *Periodontol 2000* 2016;**70**(1):80−92.

57. Madsen J, Mollenhauer J, Holmskov U. Review: Gp-340/DMBT1 in mucosal innate immunity. *Innate Immun* 2010;**16**(3):160−7.

58. Bajaj JS, Betrapally NS, Hylemon PB, et al. Salivary microbiota reflects changes in gut microbiota in cirrhosis with hepatic encephalopathy. *Hepatology* 2015;**62**(4):1260−71.

59. Lenander-Lumikari M, Tenovuo J, Puhakka HJ, et al. Salivary antimicrobial proteins and mutans streptococci in tonsillectomized children. *Pediatr Dent* 1992;**14**(2):86−91.

60. Baker N, Hansson GC, Leffler H, Riise G, Svanborg-Eden C. Glycosphingolipid receptors for *Pseudomonas aeruginosa*. *Infect Immun* 1990;**58**(7):2361−6.

61. Canullo L, Rossetti PH, Penarrocha D. Identification of *Enterococcus faecalis* and *Pseudomonas aeruginosa* on and in implants in individuals with peri-implant disease: a cross-sectional study. *Int J Oral Maxillofac Implants* 2015;**30**(3):583−7.

62. Paranchych W, Sastry PA, Volpel K, Loh BA, Speert DP. Fimbriae (pili): molecular basis of *Pseudomonas aeruginosa* adherence. *Clin Invest Med* 1986;**9**(2):113−18.

63. Doig P, Paranchych W, Sastry PA, Irvin RT. Human buccal epithelial cell receptors of *Pseudomonas aeruginosa*: identification of glycoproteins with pilus binding activity. *Can J Microbiol* 1989;**35**(12):1141−5.

64. Doig P, Sastry PA, Hodges RS, Lee KK, Paranchych W, Irvin RT. Inhibition of pilus-mediated adhesion of *Pseudomonas aeruginosa* to human buccal epithelial cells by monoclonal antibodies directed against pili. *Infect Immun* 1990;**58**(1):124−30.

65. Taylor CJ, McGaw J, Howden R, Duerden BI, Baxter PS. Bacterial reservoirs in cystic fibrosis. *Arch Dis Child* 1990;**65**(2):175−7.

66. Komiyama K, Habbick BF, Tumber SK. Whole, submandibular, and parotid saliva-mediated aggregation of *Pseudomonas aeruginosa* in cystic fibrosis. *Infect Immun* 1989;**57**(4):1299−304.

67. Reddy MS. Binding of the pili of *Pseudomonas aeruginosa* to a low-molecular-weight mucin and neutral cystatin of human submandibular-sublingual saliva. *Curr Microbiol* 1998;**37**(6):395−402.

68. Biesbrock AR, Reddy MS, Levine MJ. Interaction of a salivary mucin-secretory immunoglobulin A complex with mucosal pathogens. *Infect Immun* 1991;**59**(10):3492−7.

69. Komiyama K, Gibbons RJ. Interbacterial adherence between *Actinomyces viscosus* and strains of *Streptococcus pyogenes*, *Streptococcus agalactiae*, and *Pseudomonas aeruginosa*. *Infect Immun* 1984;**44**(1):86−90.

70. Uehara Y, Kikuchi K, Nakamura T, et al. H(2)O(2) produced by viridans group streptococci may contribute to inhibition of methicillin-resistant staphylococcus aureus colonization of oral cavities in newborns. *Clin Infect Dis* 2001;**32**(10):1408−13.

71. Fultz PN. Components of saliva inactivate human immunodeficiency virus. *Lancet* 1986;**2** (8517):1215.

72. Fox PC, Wolff A, Yeh CK, Atkinson JC, Baum BJ. Saliva inhibits HIV-1 infectivity. *J Am Dent Assoc* 1988;**116**(6):635−7.

73. Fox PC, Wolff A, Yeh CK, Atkinson JC, Baum BJ. Salivary inhibition of HIV-1 infectivity: functional properties and distribution in men, women, and children. *J Am Dent Assoc* 1989;**118**(6):709−11.

74. Archibald DW, Cole GA. In vitro inhibition of HIV-1 infectivity by human salivas. *AIDS Res Hum Retroviruses* 1990;**6**(12):1425−32.

75. Moore BE, Flaitz CM, Coppenhaver DH, et al. HIV recovery from saliva before and after dental treatment: inhibitors may have critical role in viral inactivation. *J Am Dent Assoc* 1993;**124**(10):67−74.

76. Robinovitch MR, Iversen JM, Resnick L. Anti-infectivity activity of human salivary secretions toward human immunodeficiency virus. *Crit Rev Oral Biol Med* 1993;**4**(3-4):455−9.

77. Baron S, Poast J, Cloyd MW. Why is HIV rarely transmitted by oral secretions? Saliva can disrupt orally shed, infected leukocytes. *Arch Intern Med* 1999;**159**(3):303–10.

78. Baron S, Poast J, Richardson CJ, Nguyen D, Cloyd M. Oral transmission of human immunodeficiency virus by infected seminal fluid and milk: a novel mechanism. *J Infect Dis* 2000;**181**(2):498–504.

79. Tebit DM, Ndembi N, Weinberg A, Quinones-Mateu ME. Mucosal transmission of human immunodeficiency virus. *Curr HIV Res* 2012;**10**(1):3–8.

80. Nittayananta W, Weinberg A, Malamud D, Moyes D, Webster-Cyriaque J, Ghosh S. Innate immunity in HIV-1 infection: epithelial and non-specific host factors of mucosal immunity- a workshop report. *Oral Dis* 2016;**22**(Suppl. 1):171–80.

81. Clapham PR, McKnight A. HIV-1 receptors and cell tropism. *Br Med Bull* 2001;**58**(1):43–59.

82. Han Y, Lai J, Barditch-Crovo P, et al. The role of protective HCP5 and HLA-C associated polymorphisms in the control of HIV-1 replication in a subset of elite suppressors. *AIDS* 2008;**22**(4):541–4.

83. Wu Z, Lee S, Abrams W, Weissman D, Malamud D. The N-terminal SRCR-SID domain of gp-340 interacts with HIV type 1 gp120 sequences and inhibits viral infection. *AIDS Res Hum Retroviruses* 2006;**22**(6):508–15.

84. Cannon G, Yi Y, Ni H, et al. HIV envelope binding by macrophage-expressed gp340 promotes HIV-1 infection. *J Immunol* 2008;**181**(3):2065–70.

85. Stoddard E, Ni H, Cannon G, et al. Gp340 promotes transcytosis of human immunodeficiency virus type 1 in genital tract-derived cell lines and primary endocervical tissue. *J Virol* 2009;**83**(17):8596–603.

86. Patyka M, Malamud D, Weissman D, Abrams WR, Kurago Z. Periluminal distribution of HIV-binding target cells and Gp340 in the oral, cervical and sigmoid/rectal mucosae: a mapping study. *PLoS One* 2015;**10**(7):e0132942.

87. Bukawa H, Sekigawa K, Hamajima K, et al. Neutralization of HIV-1 by secretory IgA induced by oral immunization with a new macromolecular multicomponent peptide vaccine candidate. *Nat Med* 1995;**1**(7):681–5.

88. Peacocke J, Lotz Z, de Beer C, Roux P, Mall AS. The role of crude saliva and purified salivary mucins in the inhibition of the human immunodeficiency virus type 1. *Virol J* 2012;**9**:177.

89. Crombie R, Silverstein RL, MacLow C, Pearce SF, Nachman RL, Laurence J. Identification of a CD36-related thrombospondin 1-binding domain in HIV-1 envelope glycoprotein gp120: relationship to HIV-1-specific inhibitory factors in human saliva. *J Exp Med* 1998;**187**(1):25–35.

90. Nagashunmugam T, Malamud D, Davis C, Abrams WR, Friedman HM. Human submandibular saliva inhibits human immunodeficiency virus type 1 infection by displacing envelope glycoprotein gp120 from the virus. *J Infect Dis* 1998;**178**(6):1635–41.

91. McNeely TB, Dealy M, Dripps DJ, Orenstein JM, Eisenberg SP, Wahl SM. Secretory leukocyte protease inhibitor: a human saliva protein exhibiting anti-human immunodeficiency virus 1 activity in vitro. *J Clin Invest* 1995;**96**(1):456–64.

92. McNeely TB, Shugars DC, Rosendahl M, Tucker C, Eisenberg SP, Wahl SM. Inhibition of human immunodeficiency virus type 1 infectivity by secretory leukocyte protease inhibitor occurs prior to viral reverse transcription. *Blood* 1997;**90**(3):1141–9.

93. Sun L, Finnegan CM, Kish-Catalone T, et al. Human beta-defensins suppress human immunodeficiency virus infection: potential role in mucosal protection. *J Virol* 2005;**79** (22):14318–29.

94. Weinberg A, Quinones-Mateu ME, Lederman MM. Role of human beta-defensins in HIV infection. *Adv Dent Res* 2006;**19**(1):42–8.

95. White MR, Helmerhorst EJ, Ligtenberg A, et al. Multiple components contribute to ability of saliva to inhibit influenza viruses. *Oral Microbiol Immunol* 2009;**24**(1):18–24.

96. Hartshorn KL, Ligtenberg A, White MR, et al. Salivary agglutinin and lung scavenger receptor cysteine-rich glycoprotein 340 have broad anti-influenza activities and interactions with surfactant protein D that vary according to donor source and sialylation. *Biochem J* 2006;**393**(Pt 2):545–53.

97. Chen CH, Zhang XQ, Lo CW, et al. The essentiality of alpha-2-macroglobulin in human salivary innate immunity against new H1N1 swine origin influenza A virus. *Proteomics* 2010;**10**(12):2396–401.

98. Qin Y, Zhong Y, Zhu M, et al. Age- and sex-associated differences in the glycopatterns of human salivary glycoproteins and their roles against influenza A virus. *J Proteome Res* 2013;**12**(6):2742–54.

99. Ueta E, Tanida T, Doi S, Osaki T. Regulation of *Candida albicans* growth and adhesion by saliva. *J Lab Clin Med* 2000;**136**(1):66–73.

100. Radfar L, Shea Y, Fischer SH, et al. Fungal load and candidiasis in Sjogren's syndrome. *Oral Surg Oral Med Oral Pathol Oral Radiol Endod* 2003;**96**(3):283–7.

101. Tanida T, Okamoto T, Okamoto A, et al. Decreased excretion of antimicrobial proteins and peptides in saliva of patients with oral candidiasis. *J Oral Pathol Med* 2003;**32**(10):586–94.

102. Nikawa H, Jin C, Makihira S, Hamada T, Samaranayake LP. Susceptibility of *Candida albicans* isolates from the oral cavities of HIV-positive patients to histatin-5. *J Prosthet Dent* 2002;**88**(3):263–7.

103. Torres SR, Garzino-Demo A, Meiller TF, Meeks V, Jabra-Rizk MA. Salivary histatin-5 and oral fungal colonisation in HIV + individuals. *Mycoses* 2009;**52**(1):11–15.

104. Fitzgerald DH, Coleman DC, O'Connell BC. Susceptibility of *Candida dubliniensis* to salivary histatin 3. *Antimicrob Agents Chemother* 2003;**47**(1):70–6.

105. Sweet SP, Denbury AN, Challacombe SJ. Salivary calprotectin levels are raised in patients with oral candidiasis or Sjogren's syndrome but decreased by HIV infection. *Oral Microbiol Immunol* 2001;**16**(2):119–23.

106. Lindh E, Brannstrom J, Jones P, et al. Autoimmunity and cystatin SA1 deficiency behind chronic mucocutaneous candidiasis in autoimmune polyendocrine syndrome type 1. *J Autoimmun* 2013;**42**:1–6.

107. Conti HR, Baker O, Freeman AF, et al. New mechanism of oral immunity to mucosal candidiasis in hyper-IgE syndrome. *Mucosal Immunol* 2011;**4**(4):448–55.

108. Lin AL, Johnson DA, Patterson TF, et al. Salivary anticandidal activity and saliva composition in an HIV-infected cohort. *Oral Microbiol Immunol* 2001;**16**(5):270–8.

109. Laibe S, Bard E, Biichle S, et al. New sensitive method for the measurement of lysozyme and lactoferrin to explore mucosal innate immunity. Part II: Time-resolved immunofluorometric assay used in HIV patients with oral candidiasis. *Clin Chem Lab Med* 2003;**41**(2):134–8.

110. Samaranayake YH, Samaranayake LP, Pow EH, Beena VT, Yeung KW. Antifungal effects of lysozyme and lactoferrin against genetically similar, sequential *Candida albicans* isolates from a human immunodeficiency virus-infected southern chinese cohort. *J Clin Microbiol* 2001;**39**(9):3296–302.

111. Bard E, Laibe S, Clair S, et al. Nonspecific secretory immunity in HIV-infected patients with oral candidiasis. *J Acquir Immune Defic Syndr* 2002;**31**(3):276–84.

112. Naglik JR, Rodgers CA, Shirlaw PJ, et al. Differential expression of *Candida albicans* secreted aspartyl proteinase and phospholipase B genes in humans correlates with active oral and vaginal infections. *J Infect Dis* 2003;**188**(3):469–79.

113. Millon L, Drobacheff C, Piarroux R, et al. Longitudinal study of anti-candida albicans mucosal immunity against aspartic proteinases in HIV-infected patients. *J Acquir Immune Defic Syndr* 2001;**26**(2):137–44.

114. Drobacheff C, Millon L, Monod M, et al. Increased serum and salivary immunoglobulins against *Candida albicans* in HIV-infected patients with oral candidiasis. *Clin Chem Lab Med* 2001;**39**(6):519–26.

115. Sawaki K, Mizukawa N, Yamaai T, Fukunaga J, Sugahara T. Immunohistochemical study on expression of alpha-defensin and beta-defensin-2 in human buccal epithelia with candidiasis. *Oral Dis* 2002;**8**(1):37–41.

116. Jurevic RJ, Bai M, Chadwick RB, White TC, Dale BA. Single-nucleotide polymorphisms (SNPs) in human beta-defensin 1: high-throughput SNP assays and association with candida carriage in type I diabetics and nondiabetic controls. *J Clin Microbiol* 2003;**41**(1):90–6.

117. Elahi S, Pang G, Ashman RB, Clancy R. Nitric oxide-enhanced resistance to oral candidiasis. *Immunology* 2001;**104**(4):447–54.

118. Steele C, Leigh J, Slobodan R, Fidel Jr. PL. Growth inhibition of candida by human oral epithelial cells. *J Infect Dis* 2000;**182**(5):1479–85.

119. Fidel Jr. PL. Immunity to candida. *Oral Dis* 2002;**8**(Suppl. 2):69–75.

120. Quimby K, Lilly EA, Zacharek M, et al. CD8 T cells and E-cadherin in host responses against oropharyngeal candidiasis. *Oral Dis* 2012;**18**(2):153–61.

121. Bishu S, Su EW, Wilkerson ER, et al. Rheumatoid arthritis patients exhibit impaired *Candida albicans*-specific Th17 responses. *Arthritis Res Ther* 2014;**16**(1):R50.

122. Oliveira MA, Carvalho LP, Gomes Mde S, Bacellar O, Barros TF, Carvalho EM. Microbiological and immunological features of oral candidiasis. *Microbiol Immunol* 2007;**51**(8):713–19.

123. Conti HR, Shen F, Nayyar N, et al. Th17 cells and IL-17 receptor signaling are essential for mucosal host defense against oral candidiasis. *J Exp Med* 2009;**206**(2):299–311.

124. Tardif F, Goulet JP, Zakrazewski A, Chauvin P, Rouabhia M. Involvement of interleukin-18 in the inflammatory response against oropharyngeal candidiasis. *Med Sci Monit* 2004;**10**(8):BR239–49.

125. Thein ZM, Samaranayake YH, Samaranayake LP. Characteristics of dual species candida biofilms on denture acrylic surfaces. *Arch Oral Biol* 2007;**52**(12):1200–8.

126. Jin Y, Samaranayake LP, Samaranayake Y, Yip HK. Biofilm formation of *Candida albicans* is variably affected by saliva and dietary sugars. *Arch Oral Biol* 2004;**49**(10):789–98.

127. Naglik JR, Fostira F, Ruprai J, Staab JF, Challacombe SJ, Sundstrom P. *Candida albicans* HWP1 gene expression and host antibody responses in colonization and disease. *J Med Microbiol* 2006;**55**(Pt 10):1323–7.

128. Stecksen-Blicks C, Holgerson PL, Twetman S. Effect of xylitol and xylitol-fluoride lozenges on approximal caries development in high-caries-risk children. *Int J Paediatr Dent* 2008;**18**(3):170–7.

129. Singh RP, Damle SG, Chawla A. Salivary mutans streptococci and lactobacilli modulations in young children on consumption of probiotic ice-cream containing *Bifidobacterium lactis* Bb12 and lactobacillus acidophilus La5. *Acta Odontol Scand* 2011;**69**(6):389–94.

CHAPTER 2

Oral Bacteria Transmissible by Saliva and Kissing

Saliva contains microorganisms, particularly various bacteria, derived mainly from exfoliation from oral surfaces,[1] especially the dorsum of the tongue, and from other locations—notably the respiratory tract and bleeding onto the mouth. Identification of oral bacteria may be confirmed by culture but approximately 30% are known as uncultivated phylotypes. Bacteria are identified routinely by morphological and biochemical tests, but *nucleic acid*–based methods are the most rapid and specific tools. Kissing often implies an exchange of saliva, making it an activity that can potentially produce transmission of oral bacteria from person-to-person until a susceptible recipient is reached who will develop the infection.

Bacteria are generally regarded as harmful, and virulence factors appear important in colonization, such as the mutacin—a bacteriocin-like inhibitory substance—of *Streptococcus mutans*.[2] However, not all bacteria are necessarily harmful, and some may have beneficial effects; e.g., salivary microbiota may have protective effects against colonization by opportunistic pathogens such as *Pseudomonas aeruginosa*.[3]

2.1 DEVELOPMENT OF THE ORAL MICROBIOME

The oral cavity is the portal for initial entry of oral and gut indigenous colonization. Neonates generally lack oral cultivatable bacteria but, by 48 hours, these are detectable, and there is a significant correlation between the bacterial counts of neonates and those of their mothers.[4] At birth, the type of delivery influences the early salivary flora[5] and thereafter, skin-to-skin care, in which the mother holds the naked infant in a nappy (diaper) between her breasts, is associated with a distinct oral microbial flora—with increased *Streptococci*.[6] There is transmission of cariogenic bacteria within families.[7–10] Beyond the neonatal period, all healthy individuals develop an oral bacterial flora "normal" for them.[11] In infant saliva the predominant genera is *Streptococcus* along with *Veillonella*,

Saliva Protection and Transmissible Diseases. DOI: http://dx.doi.org/10.1016/B978-0-12-813681-2.00002-0
© 2017 Elsevier Inc. All rights reserved.

Neisseria, Rothia, Haemophilus, Gemella, Granulicatella, Leptotrichia, and *Fusobacterium; Firmicutes, Proteobacteria, Actinobacteria,* and other *Fusobacteria* are also present.[12]

2.2 THE ADULT SALIVARY MICROBIOME

By adulthood, *Actinomyces, Fusobacterium, Haemophilus, Neisseria, Oribacterium, Rothia, Treponema,* and *Veillonella,* predominate, along with *Firmicutes, Proteobacteria, Actinobacteria,* and other *Fusobacteria.*[12] The oral microbiome of systemically healthy adults may include *Actinomyces, Capnocytophaga, Escherichia, Gemella, Granulicatella, Haemophilus, Neisseria, Oribacterium, Peptostreptococcus, Prevotella, Streptococcus, Veillonella*—and nonculturable bacteria.[13] Human Oral Microbe Identification Microarray assays show that *Streptococcus* and *Veillonella* are the most predominant genera identified, and that age, gender, diet, alcohol consumption, and body mass index have minimal effect on the bacterial profile, but smoking and socioeconomic status can affect the flora.[14] Various cocci, mycobacteria, and treponemes may also be found in saliva, in quantities possibly sufficient to infect other individuals.[15]

The salivary microbiome thus becomes complex and composed of bacteria from different oral surfaces.[16] Cariogenic and periodontopathogenic bacterial flora habitually colonize the human mouth and this literature is well reviewed elsewhere.[17]

There is little or no variation in microbial profiles over time[18] and similarity between siblings seems to be retained,[19] but oral hygiene affects the bacterial diversity in saliva,[13] as does oral disease.[20]

2.3 EFFECTS OF DISEASE ON THE SALIVARY MICROBIOME

Different salivary bacterial profiles are seen in oral health and disease[21] and the use of nucleic acid sequencing, checkerboard DNA–DNA hybridization, pyrosequencing, and Illumina high-throughput sequencing is now allowing better characterization of the microbiome.[22–25] Most oral *Streptococcus mutans, Porphyromonas gingivalis, Tannerella forsythia,* and *Treponema denticola* strains are not culturable and it is increasingly clear that the oral microbiome is more diverse than originally believed and that disease is polymicrobial.[22,26,27]

The salivary bacterial flora differs where there is oral disease such as periodontitis and dental caries,[28] and the microbial profiles of saliva can differentiate between patients with dental caries, patients with periodontitis, and individuals with a healthy mouth.[28] The intraoral transmission of these bacteria seems to be facilitated by contaminated oral hygiene devices.[29]

Infections are seen especially in immunocompromised people and can be life-changing or even lethal. For example, the immunosuppression aimed at inhibiting T-cell-mediated responses creates a permissive oral environment for opportunistic pathogens in organ transplant recipients but has little effect on the rest of the salivary bacteriome.[30] The taxa more prevalent in the mouth of immunocompromised patients than in controls may include operational taxonomic units (OTUs) of potentially opportunistic Gammaproteobacteria (e.g., *Klebsiella pneumoniae*, *Pseudomonas fluorescens*, *Acinetobacter* spp., *Vibrio* spp., *Enterobacteriaceae* spp., and the genera *Acinetobacter* and *Klebsiella*). Transplant subjects also have increased proportions in the mouth of *Pseudomonas aeruginosa*, *Acinetobacter* spp., *Enterobacteriaceae* spp., and *Enterococcus faecalis*, among other OTUs, while genera with increased proportions included *Klebsiella*, *Acinetobacter*, *Staphylococcus*, and *Enterococcus*.[30] The oral microbiomes of HIV-positive and HIV-negative individuals were found to be similar overall, although an OTU identified as *Haemophilus parainfluenzae* is significantly associated with HIV patients while *Streptococcus mitis*/HOT473 was most significantly associated with HIV-negative individuals.[31] The main salivary bacteria in HIV/AIDS patients include *Firmicutes*, *Bacteroidetes*, and *Proteobacteria*, *Capnocytophaga* spp., and others such as *Mycoplasma salivarium*, *Neisseria elongata*, and *Streptococcus mitis* may be opportunistic.[32] Variations in salivary microbiota have also been implicated in the etiology of some cancers. (e.g., *N. elongata* and *S. mitis* are usually found in saliva of pancreatic cancer cases.)[33]

2.4 SALIVA AND INFECTION TRANSMISSION

The specific effects of kissing on the transmission of salivary microbiota are unknown. Although no quantitative model is available that enables us to analyze the dynamics of the transitory and permanent bacteria involved, there appear to be no significant variation in the microbiota of the saliva (or tongue) after a single intimate kiss.[34] After administering yogurts containing bacteria (*Lactobacillus* and *Bifidobacteria*) as a test to a group of healthy volunteers, it was estimated a mean bacterial load transfer of 0.8×10^8 bacteria per 10-second intimate kiss. The results of this study suggested that transmitted bacteria were present only transiently, whereas the tongue surface favored long-term bacterial colonization.[34]

Most evidence of infection transmission comes from epidemiological studies but even then it may be difficult to differentiate saliva transmission from that by fomites or body fluids introduced into the mouth, respiratory droplets, aerosols, or other routes. For example, hospitalized patients may have respiratory pathogens in their saliva,[35] and this is especially the case in mechanically ventilated patients.[36]

2.5 EFFECTS OF HYPOSALIVATION

Hyposalivation may result in increases in the oral flora and in related diseases.[37] For example, in patients undergoing radiotherapy treatment for head and neck cancer the most frequently isolated bacteria were *Citrobacter*, *Enterobacter*, *Enterococcus*, *Klebsiella*, *Proteus*, and *Pseudomonas*.[38] However, the salivary profiles of caries or periodontitis-associated bacteria do not differ in patients with severe hyposalivation from those with normal salivation unless there is active dental caries.[39]

2.6 DENTAL CARIES TRANSMISSION

The question of dental caries and transmission of responsible bacteria is pertinent. *Streptococcus mutans* and *Streptococcus sobrinus* are the most prevalent caries-associated microorganisms,[40] and *Lactobacilli* are a major contributor to caries progression.[41] *Actinomyces*, *Bifidobacteriaceae*, *Granulicatella*, *Scardovia*, and *Veillonella* have also been identified in early childhood caries.[42-45] *Solobacterium moorei*, *Streptococcus salivarius*, and *Streptococcus parasanguinis* are found at higher levels in saliva from adults with caries.[46]

Newborn infants generally lack the cariogenic oral *S. mutans* and aerobic cultivatable bacteria but, by 48 hours, these are detectable, and there is a significant correlation between the bacterial counts of neonates and those of their mothers.[4] There is a higher prevalence of *S. mutans* in full-term infants than in premature infants and often transmission of cariogenic bacteria (*S. mutans*, *Lactobacillus* spp., and *Actinomyces* spp.) to the oral cavity by 1 year—when cariogenic microbes are present in most premature very low birthweight infants and in all full-term infants.[47] Genotyping studies strongly suggest that *S. mutans* spreads vertically in the population, mostly from mothers to their children[48] and, between ages 19 and 31 months, many children have acquired *S. mutans* from their mothers.[49]

Saliva is the main vehicle by which this transfer occurs.[8] The transmission rate for *S. mutans* is higher in mothers with saliva levels $>10^6$ microorganisms per mL.[7] Mother-to-child transmission of *Lactobacillus* has also been demonstrated, again particularly when the maternal salivary bacterial load is high.[10] Furthermore, a reduction in the maternal oral *S. mutans* load delays and reduces transmission to their offspring.[9]

The intraoral transmission of cariogenic species seems to be also facilitated by contaminated toothbrushes and other oral hygiene devices. The contaminating flora decreases after dry storage, but a small number still detectable on toothbrushes by 24 hours. Toothpaste reduces the microbial load after 24 hours of storage.[29]

One of the main routes of transmission of cariogenic bacteria appears to be direct contact, often in kissing[50]; frequent kissing on the lips seems one of the most significant factors associated with *S. mutans* colonization.[51]

2.7 PERIODONTAL DISEASE TRANSMISSION

Periodontal pathogens may include *Aggregatibacter actinomycetemcomitans*, *Fusobacterium nucleatum*, *Peptostreptococcus micros*, *Pophyromonas gingivalis*, *Prevotella intermedia*, *Treponema denticola*, *Treponema forsythia*, and putative periodontal pathogens such as *Filifactor alocis* and *Parvimonas micra*[52]—especially organisms of the red complex (*P. gingivalis*, *T. forsythia* and *T. denticola*)—though at least 17 new additional candidate organisms, including species or phylotypes from the phyla *Bacteroidetes*, *Candidatus Saccharibacteria*, *Firmicutes*, *Proteobacteria*, *Spirochaetes*, and *Synergistetes*, as well as possibly from the Archaea domain are reportedly associated with periodontal disease.[53]

In one study, periodontal pathogens were present in 1-year olds in 83% of preterm infants and in 96% of full-term infants.[47] Putative periodontal pathogens may spread between family members[54] which could explain how colonization by periodontal pathogens in early childhood occurs in a larger proportion than previously reported.[55] A marked correlation in the presence of periodontal pathogens, particularly the red complex species was found between mothers and their children.[56] Vertical transmission of *A. actinomycetemcomitans* between mothers and children ranges from 30% to 60%.[57] Mothers colonized by *P. gingivalis* are a risk factor for oral colonization of their children and the risk is greatest when both parents are colonized.[58] Likewise, oral *T. forsythia*, *P. Intermedia*, and *P. nigrescens* are more frequently found in children whose parents harbored these bacteria.[59]

The horizontal transmission of *A. actinomycetemcomitans* between couples has been estimated to be from 14% to 60%, and that of *P. gingivalis* from 30% to 75%.[57] Many periodontopathic bacteria are detectable in saliva even after removal of all teeth,[60] suggesting that saliva is a vehicle of person-to-person transmission of these bacteria.[61]

Saliva transmission of periodontopathogens via kissing is highly likely but none of these studies absolutely confirmed kissing as the source of transmission which might also, e.g., be spread via fomites.

2.8 ACUTE NECROTIZING GINGIVITIS AND PERIODONTITIS

Studies of the microbiome in acute necrotizing gingivitis (ANUG) are few. ANUG may rarely progress to gangrene (noma or cancrum oris), mainly where there is an immune defect such as in malnourishment or HIV/AIDS. In ANUG, the main oral bacteria isolated include *P. intermedia*, α-hemolytic streptococci, and *Actinomyces* spp., and noma is associated with large numbers of fusiform bacilli and spirochaetes.[62] Patients with necrotic periodontitis in HIV/AIDS have a distinct salivary bacterial profile from those with chronic periodontitis; *Dialister pneumosintes*, *Eubacterium infirmum*, *Rothia mucilaginosa*, and *Treponema parvum* are preponderant in AIDS patients with periodontitis.[32]

There appears to be no reliable evidence that ANUG is transmitted by saliva or by kissing but it must be a possibility.

2.9 CLOSING REMARKS AND PERSPECTIVES

Apart from the cariogenic and periodontopathogenic flora, a considerable number of pathogenic bacteria responsible for medical diseases can be present in saliva and may be transmitted to individuals in close contact with the host. Although the majority of these pathogens are mostly only detected in saliva during the acute phase of infection, microorganism shedding can also occur in the saliva of clinically healthy subjects. The bacterial load is usually lower in saliva than in other body fluids, such as blood, but it can be sufficient to transmit disease. Based on epidemiological studies, it can be concluded that kissing is a route of transmission of oral bacteria. Rigorous studies using molecular biology techniques are needed to establish genetic homology between the bacteria isolated from saliva samples from kissing partners and more definitively establish which microorganisms can be transmitted by kissing but the evidence suggests bacteria associated with the main oral diseases can be transmitted.

REFERENCES

1. Segata N, Haake SK, Mannon P, et al. Composition of the adult digestive tract bacterial microbiome based on seven mouth surfaces, tonsils, throat and stool samples. *Genome Biol* 2012;**13**(6):R42.

2. Gronroos L, Saarela M, Matto J, Tanner-Salo U, Vuorela A, Alaluusua S. Mutacin production by *Streptococcus mutans* may promote transmission of bacteria from mother to child. *Infect Immun* 1998;**66**(6):2595–600.

3. He X, Hu W, He J, Guo L, Lux R, Shi W. Community-based interference against integration of *Pseudomonas aeruginosa* into human salivary microbial biofilm. *Mol Oral Microbiol* 2011;**26**(6):337–52.

4. Rosenblatt R, Steinberg D, Mankuta D, Zini A. Acquired oral microflora of newborns during the first 48 hours of life. *J Clin Pediatr Dent* 2015;**39**(5):442–6.

5. Dominguez-Bello MG, Costello EK, Contreras M, et al. Delivery mode shapes the acquisition and structure of the initial microbiota across multiple body habitats in newborns. *Proc Natl Acad Sci USA* 2010;**107**(26):11971–5.

6. Hendricks-Munoz KD, Xu J, Parikh HI, et al. Skin-to-skin care and the development of the preterm infant oral microbiome. *Am J Perinatol* 2015;**32**(13):1205–16.

7. Berkowitz RJ, Turner J, Green P. Maternal salivary levels of streptococcus mutans and primary oral infection of infants. *Arch Oral Biol* 1981;**26**(2):147–9.

8. Chan KM, King NM, Kilpatri NM. Can infants catch caries? A review of the current evidence on the infectious nature of dental caries in infants. *N Z Dent J* 2005;**101**(1):4–11.

9. Nakai Y, Shinga-Ishihara C, Kaji M, Moriya K, Murakami-Yamanaka K, Takimura M. Xylitol gum and maternal transmission of mutans streptococci. *J Dent Res* 2010;**89** (1):56–60.

10. Teanpaisan R, Chaethong W, Piwat S, Thitasomakul S. Vertical transmission of mutans streptococci and lactobacillus in Thai families. *Pediatr Dent* 2012;**34**(2):e24–9.

11. Huse SM, Ye Y, Zhou Y, Fodor AA. A core human microbiome as viewed through 16S rRNA sequence clusters. *PLoS One* 2012;**7**(6):e34242.

12. Cephas KD, Kim J, Mathai RA, et al. Comparative analysis of salivary bacterial microbiome diversity in edentulous infants and their mothers or primary care givers using pyrosequencing. *PLoS One* 2011;**6**(8):e23503.

13. Pereira JV, Leomil L, Rodrigues-Albuquerque F, Pereira JO, Astolfi-Filho S. Bacterial diversity in the saliva of patients with different oral hygiene indexes. *Braz Dent J* 2012;**23**(4):409–16.

14. Belstrom D, Holmstrup P, Nielsen CH, et al. Bacterial profiles of saliva in relation to diet, lifestyle factors, and socioeconomic status. *J Oral Microbiol* 2014;6.

15. Kang JG, Kim SH, Ahn TY. Bacterial diversity in the human saliva from different ages. *J Microbiol* 2006;**44**(5):572–6.

16. Samaranayake L. Normal oral flora, the oral ecosystem and plaque biofilms. In: Samaranayake L, editor. *Essential microbiology for dentistry*. 4th ed. London, UK: Churchill Livingstone; 2011. p. 265–78.

17. Marsh P, Martin MV. *Oral microbiology*. 4th ed. Oxford, UK: Wright, Butterworth-Heinemann; 1999.

18. Belstrom D, Holmstrup P, Bardow A, Kokaras A, Fiehn NE, Paster BJ. Temporal stability of the salivary microbiota in oral health. *PLoS One* 2016;**11**(1):e0147472.

19. Lee HS, Lee JH, Kim SO, et al. Comparison of the oral microbiome of siblings using next-generation sequencing: a pilot study. *Oral Dis* 2016;**22**(6):549–56.

20. Pozhitkov AE, Leroux BG, Randolph TW, Beikler T, Flemmig TF, Noble PA. Towards microbiome transplant as a therapy for periodontitis: an exploratory study of periodontitis microbial signature contrasted by oral health, caries and edentulism. *BMC Oral Health* 2015;**15**:125.

21. Belstrom D, Paster BJ, Fiehn NE, Bardow A, Holmstrup P. Salivary bacterial fingerprints of established oral disease revealed by the human oral microbe identification using next generation sequencing (HOMINGS) technique. *J Oral Microbiol* 2016;8.

22. Nasidze I, Li J, Quinque D, Tang K, Stoneking M. Global diversity in the human salivary microbiome. *Genome Res* 2009;**19**(4):636–43.

23. Nasidze I, Quinque D, Li J, Li M, Tang K, Stoneking M. Comparative analysis of human saliva microbiome diversity by barcoded pyrosequencing and cloning approaches. *Anal Biochem* 2009;**391**(1):64–8.

24. Aas JA, Griffen AL, Dardis SR, et al. Bacteria of dental caries in primary and permanent teeth in children and young adults. *J Clin Microbiol* 2008;**46**(4):1407–17.

25. Keijser BJ, Zaura E, Huse SM, et al. Pyrosequencing analysis of the oral microflora of healthy adults. *J Dent Res* 2008;**87**(11):1016–20.

26. Zaura E, Keijser BJ, Huse SM, Crielaard W. Defining the healthy "core microbiome" of oral microbial communities. *BMC Microbiol* 2009;**9**:259.

27. Ling Z, Kong J, Jia P, et al. Analysis of oral microbiota in children with dental caries by PCR-DGGE and barcoded pyrosequencing. *Microb Ecol* 2010;**60**(3):677–90.

28. Belstrom D, Fiehn NE, Nielsen CH, et al. Differentiation of salivary bacterial profiles of subjects with periodontitis and dental caries. *J Oral Microbiol* 2015;7.

29. Schmidt JC, Bux M, Filipuzzi-Jenny E, et al. Influence of time, toothpaste and saliva in the retention of *Streptococcus mutans* and *Streptococcus sanguinis* on different toothbrushes. *J Appl Oral Sci* 2014;**22**(3):152–8.

30. Diaz PI, Hong BY, Frias-Lopez J, et al. Transplantation-associated long-term immunosuppression promotes oral colonization by potentially opportunistic pathogens without impacting other members of the salivary bacteriome. *Clin Vaccine Immunol* 2013;**20**(6):920–30.

31. Kistler JO, Arirachakaran P, Poovorawan Y, Dahlen G, Wade WG. The oral microbiome in human immunodeficiency virus (HIV)-positive individuals. *J Med Microbiol* 2015;**64**(9):1094–101.

32. Zhang F, He S, Jin J, Dong G, Wu H. Exploring salivary microbiota in AIDS patients with different periodontal statuses using 454 GS-FLX titanium pyrosequencing. *Front Cell Infect Microbiol* 2015;**5**:55.

33. Farrell JJ, Zhang L, Zhou H, et al. Variations of oral microbiota are associated with pancreatic diseases including pancreatic cancer. *Gut* 2012;**61**(4):582–8.

34. Kort R, Caspers M, van de Graaf A, van Egmond W, Keijser B, Roeselers G. Shaping the oral microbiota through intimate kissing. *Microbiome* 2014;**2**:41.

35. Zuanazzi D, Souto R, Mattos MB, et al. Prevalence of potential bacterial respiratory pathogens in the oral cavity of hospitalised individuals. *Arch Oral Biol* 2010;**55**(1):21–8.

36. Fourrier F, Dubois D, Pronnier P, et al. Effect of gingival and dental plaque antiseptic decontamination on nosocomial infections acquired in the intensive care unit: a double-blind placebo-controlled multicenter study. *Crit Care Med* 2005;**33**(8):1728–35.

37. DePaola DP. Saliva: the precious body fluid. *J Am Dent Assoc* 2008;**139**(Suppl.):5S–6S.

38. Gaetti-Jardim EJ, Ciesielski FI, de Sousa FR, Nwaokorie F, Schweitzer CM, Avila-Campos MJ. Occurrence of yeasts, pseudomonads and enteric bacteria in the oral cavity of patients undergoing head and neck radiotherapy. *Braz J Microbiol* 2011;**42**(3):1047–55.

39. Belstrom D, Holmstrup P, Fiehn NE, et al. Bacterial composition in whole saliva from patients with severe hyposalivation—a case-control study. *Oral Dis* 2016;**22**(4):330–7.

40. Caufield PW. Dental caries—a transmissible and infectious disease revisited: a position paper. *Pediatr Dent* 1997;**19**(8):491–8.

41. Caufield PW, Schon CN, Saraithong P, Li Y, Argimon S. Oral lactobacilli and dental caries: a model for niche adaptation in humans. *J Dent Res* 2015;**94**(9 Suppl.):110S–18S.

42. Mantzourani M, Gilbert SC, Sulong HN, et al. The isolation of bifidobacteria from occlusal carious lesions in children and adults. *Caries Res* 2009;**43**(4):308–13.

43. Kanasi E, Dewhirst FE, Chalmers NI, et al. Clonal analysis of the microbiota of severe early childhood caries. *Caries Res* 2010;**44**(5):485–97.

44. Tanner AC, Mathney JM, Kent RL, et al. Cultivable anaerobic microbiota of severe early childhood caries. *J Clin Microbiol* 2011;**49**(4):1464–74.

45. Jiang W, Zhang J, Chen H. Pyrosequencing analysis of oral microbiota in children with severe early childhood dental caries. *Curr Microbiol* 2013;**67**(5):537–42.

46. Belstrom D, Fiehn NE, Nielsen CH, et al. Altered bacterial profiles in saliva from adults with caries lesions: a case-cohort study. *Caries Res* 2014;**48**(5):368–75.

47. Merglova V, Koberova-Ivancakova R, Broukal Z, Dort J. The presence of cariogenic and periodontal pathogens in the oral cavity of one-year-old infants delivered pre-term with very low birthweights: a case control study. *BMC Oral Health* 2014;**14**:109.

48. Lapirattanakul J, Nakano K, Nomura R, Hamada S, Nakagawa I, Ooshima T. Demonstration of mother-to-child transmission of *Streptococcus mutans* using multilocus sequence typing. *Caries Res* 2008;**42**(6):466–74.

49. van Loveren C, Buijs JF, ten Cate JM. Similarity of bacteriocin activity profiles of mutans streptococci within the family when the children acquire the strains after the age of 5. *Caries Res* 2000;**34**(6):481–5.

50. Newbrun E. Preventing dental caries: breaking the chain of transmission. *J Am Dent Assoc* 1992;**123**(6):55–9.

51. Wan AK, Seow WK, Purdie DM, Bird PS, Walsh LJ, Tudehope DI. A longitudinal study of *Streptococcus mutans* colonization in infants after tooth eruption. *J Dent Res* 2003;**82** (7):504–8.

52. Belstrom D, Fiehn NE, Nielsen CH, et al. Differences in bacterial saliva profile between periodontitis patients and a control cohort. *J Clin Periodontol* 2014;**41**(2):104–12.

53. Perez-Chaparro PJ, Goncalves C, Figueiredo LC, et al. Newly identified pathogens associated with periodontitis: a systematic review. *J Dent Res* 2014;**93**(9):846–58.

54. Okada M, Hayashi F, Soda Y, Zhong X, Miura K, Kozai K. Intra-familial distribution of nine putative periodontopathogens in dental plaque samples analyzed by PCR. *J Oral Sci* 2004;**46**(3):149–56.

55. Yang EY, Tanner AC, Milgrom P, et al. Periodontal pathogen detection in gingiva/tooth and tongue flora samples from 18- to 48-month-old children and periodontal status of their mothers. *Oral Microbiol Immunol* 2002;**17**(1):55–9.

56. Tamura K, Nakano K, Hayashibara T, et al. Distribution of 10 periodontal bacteria in saliva samples from Japanese children and their mothers. *Arch Oral Biol* 2006;**51**(5):371–7.

57. Van Winkelhoff AJ, Boutaga K. Transmission of periodontal bacteria and models of infection. *J Clin Periodontol* 2005;**32**(Suppl. 6):16–27.

58. Tuite-McDonnell M, Griffen AL, Moeschberger ML, Dalton RE, Fuerst PA, Leys EJ. Concordance of *Porphyromonas gingivalis* colonization in families. *J Clin Microbiol* 1997;**35** (2):455–61.

59. Umeda M, Miwa Z, Takeuchi Y, et al. The distribution of periodontopathic bacteria among Japanese children and their parents. *J Periodontal Res* 2004;**39**(6):398–404.

60. Van Assche N, Van Essche M, Pauwels M, Teughels W, Quirynen M. Do periodontopathogens disappear after full-mouth tooth extraction?. *J Clin Periodontol* 2009;**36**(12):1043–7.

61. Asikainen S, Chen C, Slots J. Likelihood of transmitting *Actinobacillus actinomycetemcomitans* and *Porphyromonas gingivalis* in families with periodontitis. *Oral Microbiol Immunol* 1996;**11**(6):387–94.

62. Falkler Jr WA, Enwonwu CO, Idigbe EO. Microbiological understandings and mysteries of noma (*Cancrum oris*). *Oral Dis* 1999;**5**(2):150–5.

CHAPTER 3

Systemic Bacteria Transmissible by Kissing

Bacterial infections can be transmitted by various routes. At one extreme, airborne or droplet bacterial infections (e.g., tuberculosis) are fairly contagious while sexually transmitted bacterial diseases, such as syphilis, usually are only transmitted by close mucosal contacts. Saliva might contain bacteria transmissible by bites, mouth-to-mouth resuscitation, or kissing.

Human bites have transmitted a range of infections, typically polymicrobial aerobic and anaerobic infections, especially with *Eikenella, Staphylococcus*, and *Streptococcus*.[1] Human bites can also transmit viral infections such as HIV, hepatitis B, or hepatitis C.

Mouth-to-mouth resuscitation may also transmit bacterial infections, including *Helicobacter pylori, Mycobacterium tuberculosis, Neisseria meningitidis, Salmonella infantis*, and *Shigella sonnei*.[2,3]

Although the evidence is sparse, it is difficult to imagine anyone participating in intimate kisses with someone known to suffer, e.g., from syphilis or tuberculosis. In sexual activities, there is usually also kissing, especially deep (intimate, passionate or French) kissing, and there may be oro-anogenital contact that has been implicated as a route of transmission for gonorrhea, syphilis, *Chlamydia trachomatis*, chancroid, and *Neisseria meningitidis*, as well as other respiratory organisms including *Streptococci, Haemophilus influenzae*, and *Mycoplasma pneumoniae*.[4]

3.1 MEDICAL BACTERIA IN SALIVA

A considerable number of pathogenic bacteria can be responsible for medical diseases (Table 3.1) and some may be present in the saliva and potentially be conveyed especially to individuals in close contact (Table 3.2). Many of these bacteria can cause serious disease (e.g., syphilis, tuberculosis, and meningococcal infections).

Saliva Protection and Transmissible Diseases. DOI: http://dx.doi.org/10.1016/B978-0-12-813681-2.00003-2
© 2017 Elsevier Inc. All rights reserved.

Table 3.1 Main Relevant Pathogenic Medical Bacteria

Acinetobacter	*Legionella*
Actinomyces	*Mycobacteria*
Bartonella	*Pseudomonads*
Campylobacter	*Spirochetes*
Enterobacteriaceae	*Staphylococci*
Gonococci	*Streptococci*

Table 3.2 Main Medical Bacteria Detectable in Saliva (Alphabetical Order)

Bacteria	Patients With Clinical Disease	Healthy/Asymptomatic Populations
Acinetobacter spp.	NAI	Souto et al. (2014)[5]
Campylobacter	Mahendran et al. (2013)[6]	Zhang et al. (2010)[7]
Dialister pneumosintes	Zhang et al. (2015)[8]	Ferraro et al. (2007)[9]
Enterobacteriaceae	Bajaj et al. (2015)[10]	Reddy et al. (2013)[11]
Haemophilus influenzae	Foweraker et al. (1993)[12]	Hasan et al. (2014)[13]
Helicobacter pylori	Ferguson et al. (1993)[14]	Parsonnet et al. (1999)[15]
Mycobacterium tuberculosis	González Mediero et al. (2015)[16]	Palakuru et al. (2012)[17]
Mycobacterium leprae	Rosa et al. (2013)[18]	NAI
Neisseria meningitidis	NAI	MacLennan et al. (2006)[19]
Neisseria gonorrhoeae	Chow et al. (2016)[20]	NAI
Pseudomonas spp.	NAI	Souto et al. (2014)[5]
Staphylococcus spp.	NAI	Ogawa et al. (2012)[21]
Streptococcus pyogenes	Dahlén (2009)[22]	Tagg et al. (1990)[23]
Tropheryma whippelii	Fenollar et al. (2014)[24]	Dutly et al. (2000)[25]
NAI, no available information.		

However, evidence for person-to-person transmission by kissing is perhaps surprisingly, limited to a few bacterial genera and even this assertion is based only on weak scientific evidence.

3.1.1 Acinetobacteria

Acinetobacteria are Gram-negative coccobacilli. Some *Acinetobacteria* spp. such as *Acinetobacter baumannii* may cause pneumonia, skin, wound, urine, and other infections. Multiresistant *A. baumannii* (MRAB & MRAB-C) are particularly problematical in Intensive Care, Burns and Transplant units.[26] *Acinetobacteria* spp. may originate in the oral cavity, particularly when oral hygiene is poor and there is periodontal infection—then may be associated with *Pseudomonas aeruginosa* and, importantly, may be associated with pneumonia in hospitalized and institutionalized individuals[5,27] and in transplant patients.[28]

3.1.2 Actinomyces

Actinomyces is a genus of the *Actinobacteria* class, ubiquitous Grampositive bacteria which include *A. odontolyticus, A. oricola, A. radicidentis, A.c radingae,* and *A. slackii.* Some, such as *A. naeslundii, A. viscosus,* and *A. odontolyticus* may be involved in the etiopathogenesis of dental caries and periodontal disease. *Actinomyces* spp. now classified as *Aggregatibacter actinomycetemcomitans* (AA) are periodontal pathogens, which may be transmitted in saliva. *A. actinomycetemcomitans* is also a part of the *Haemophilus aphrophilus, Cardiobacterium hominis, Eikenella corrodens,* and *Kingella kingae* group of bacteria (HACEK) that cause some infective endocarditis. *A. actinomycetemcomitans* may also play a role in acute coronary syndromes,[29] and patients with diabetic kidney disease have more *A. actinomycetemcomitans* in dental plaque (and saliva) and a raised incidence of cerebral infarction.[30]

Actinomycosis is often a polymicrobial aerobic–anaerobic infection caused mainly by *Actinomyces* spp., typically by *Actinomyces israelii* or sometimes *Actinomyces gerencseriae,* and it can also be caused by *Propionibacterium propionicus* and may be associated with *A. odontolyticus* and *A. actinomycetemcomitans,* anaerobic streptococci, *E. corrodens,* and *Fusobacterium* and *Bacteroides* spp., which appear to facilitate infection by establishing a microaerophilic environment.[31] Suppuration and granulomatous inflammation lead to multiple abscesses and sinus tracts in the cervicofacial, abdominal, or pelvic areas that may discharge so-called "sulfur granules" of microbial colonies. *Actinomyces* colonies accompanied by signs of tissue invasion and reactive inflammation are a common finding in bone affected by bisphosphonate-related osteonecrosis of the jaw.[32] Actinomycosis is also a rare cause of bacterial sialadenitis.[33]

However, salivary transmission of Actinomycetes though highly probable has not been reported.

Tropheryma whippelii, closely related to the actinomycetes, causes Whipple's disease—a rare chronic multisystemic disease affecting people with defective T-lymphocyte.[34] *T. whippelii* are ubiquitous organisms found in the saliva of people with Whipple's disease,[24] and often in those without.[25,35]

T. whippelii is probably transmitted between humans via oro-oral and feco-oral routes.[24]

3.1.3 Bartonella

Bartonella (formerly *Rochalimea*) species typically cause cat-scratch fever—a common zoonosis—mainly infection with *Bartonella henselae*, transmitted from cats to humans by bites, fleas, or ticks. It typically presents as fever and tender lymphadenopathy and rarely may dissemi- nate with hepatosplenomegaly or meningoencephalitis or, in patients with HIV/AIDS may cause epithelioid (bacillary) angiomatosis.[36]

Although in humans transmission generally occurs through arthro- pod or animal bites and scratches, *B. henselae* nucleic acids have been identified in the blood and in saliva from a patient with cat-scratch fever, and his dog.[37]

Salivary transmission between humans is improbable.

3.1.4 Campylobacter

Campylobacteriosis is a common food poisoning illness. *Campylobacter concisus*, a Gram-negative, asaccharolytic, rod-shaped bacteria was originally described linked with gingivitis and periodontitis.[38] *C. concisus* has subsequently been more associated with other condi- tions, particularly Barrett's esophagus[39] and *Campylobacter* spp. are increasingly implicated in inflammatory bowel diseases,[40] *C. concisus*, being one putative such organism.[41]

The saliva may contain *C. concisus* in patients with inflammatory bowel disease[6] as well as in healthy individuals[7] but there is as yet no hard evidence for salivary transmission.

3.1.5 Chlamydia

Chlamydia trachomatis is a sexually transmitted bacterial infection, often producing no symptoms, but may cause genital discharges and may spread to the testicles or pelvis. In resource-poor countries, it has been associated with trachoma and blindness. *Chlamydia* can also be transmitted by fomites or personal contact but oral–oral transmission via saliva has rarely been reported, although oral wash specimens are more effective to detect pharyngeal *C. trachomatis* infection than pharyngeal swabs.[42] Saliva may have antichlamydial qualities.[43,44]

3.1.6 Dialister pneumosintes

Dialister pneumosintes is a nonfermentative, anaerobic, Gram-negative rod that has been recovered from gingivitis and more especially from subgingival plaque and deep periodontal pockets,[45–47] and has pathogenic potential in dental root canals, and various body sites, including the lung and brain,[27] and has been recovered from pus and body fluids and from human bite wounds.[48]

D. *pneumosintes, Eubacterium infirmum, Rothia mucilaginosa,* and *Treponema parvum* are preponderant in HIV/AIDS patients with periodontitis.[8] *D. pneumosintes* has been isolated from saliva of healthy individuals but prevalence increases in patients with periodontitis.[9]

3.1.7 Enterobacteriaceae

Enterobacteriaceae are Gram-negative bacteria of a large family that includes *Escherichia coli, Klebsiella, Salmonella, Shigella* and *Yersinia pestis.* They can cause a range of illnesses from bacteremia and endocarditis, to infections of the respiratory tract, skin, soft-tissues, urinary tract, joints, bones, eyes and CNS.

Enterobacteriaceae spp. may occasionally be isolated from saliva from apparently healthy people; most frequently detected are *Enterobacter cloacae* and *Klebsiella oxytoca.*[49] The fact that oropharyngeal *Enterobacteriaceae* isolates are unusual in normal people indicates that physical clearance and local bactericidal activity are usually effective defences.[50] However, hyposalivation and the concomitant fall in oral pH predispose to colonization with *Klebsiella pneumoniae.*[51] Hydrogen generation and/or the reduction of oxidative stress appear important for the survival and growth of *K. pneumoniae* in the oral cavity.[52]

K. pneumoniae and *E. coli* in saliva can cause pneumonia in some groups—especially in older people[21] and those who are hospitalized. The oral cavity of hospitalized patients can harbor high frequencies of *Enterobacter* spp. and other respiratory pathogens, including *Streptococci* and *Staphylococci*[53] supporting a potential for the mouth to be a reservoir for respiratory infections.[27] Several opportunistic pathogens, such as *S. pneumoniae, K. pneumoniae, E. coli,* and *S. aureus,* have been detected in saliva in patients with pneumonia, especially in older people and where the saliva is contaminated with blood.[21]

Even hospital staff may become colonized; the oral cavities of almost 20% of staff at one Oncology hospital were colonized by *Enterobacteriaceae,* over 60 different bacteria being isolated, including potentially pathogen species, the most prevalent being *Enterobacter gergoviae.*[54] *E. coli* and *Enterobacter* spp. can also be found in saliva from people who bite their nails repeatedly.[11]

Enterobacteriaceae are also increased by oral and other appliances—e.g., they are almost twice as prevalent in the oral cavities of denture-wearers as compared with others,[55] and prolonged nasogastric tube feeding is associated with colonization of the oropharynx with *P. aeruginosa* and *K. pneumoniae.*[56] Oral lesions such as oral carcinomas may also harbor significantly more bacteria including *Enterococcus faecalis, Klebsiella* spp. aerobic spore bearers, and species such as *Streptococcus, Staphylococcus, Moraxella, Citrobacter, Proteus, Pseudomonas,* as well as fungi—especially *Candida albicans.*[57]

Citrobacter, Enterobacter, Enterococcus, Klebsiella, Proteus, and Pseudomonas and yeasts (*C. albicans, C. tropicalis, C. krusei, C. glabrata* and *C. parapsilosis*) are frequently cultivatable from saliva after radiotherapy to the area, especially during mucositis.[58] There is also frequent oropharyngeal colonization of alcoholics and diabetics by Gram-negative bacilli.[59]

Apart from oral enterobacteria being more prevalent in the presence of oral appliances or lesions, or hyposalivation, and in older people or those with systemic disease, these bacteria are among the organisms colonizing immunosuppressed individuals. For example, the oral bacterial taxa more prevalent in transplant subjects include *Klebsiella pneumoniae, Pseudomonas fluorescens, Vibrio* spp., *Enterobacteriaceae* spp., and the genera *Acinetobacter* and *Klebsiella.*[28] Transplant subjects also had increased proportions of *P. aeruginosa, Acinetobacter* spp., *Enterobacteriaceae* spp., and *E. faecalis,* while genera with increased proportions included *Klebsiella, Acinetobacter, Staphylococcus,* and *Enterococcus.* But other members of the salivary bacteriome remain unaffected.[28]

A reduction in autochthonous bacteria (dysbiosis) and an increase of potentially pathogenic ones (*Enterobacteriaceae, Enterococcaceae*) is present in saliva of patients with cirrhosis, compared to controls.[10]

3.1.8 Haemophilus

Haemophilus influenzae is a Gram-negative, facultatively anaerobic coccobacillus, which can cause acute bronchitis and exacerbations of chronic obstructive pulmonary disease, as well as meningitis.

Haemophilus parainfluenzae is also a Gram-negative, facultatively anaerobic coccobacillus—part of the HACEK group—that cause about 3% of infective endocarditis cases.

Direct whole-genome shotgun metagenomic sequencing demonstrates that *Haemophilus* can be part of the salivary flora.[13] Multiple biotypes of *H. influenzae* and *H. parainfluenzae* have been isolated from saliva samples from patients with lower respiratory tract infections,[12] as well from some healthy adults,[60] though transmission by kissing has not been reported.

3.1.9 Helicobacter

Helicobacter pylori, previously *Campylobacter pylori*, is a Gram-negative, microaerophilic bacterium implicated in peptic ulceration and gastric cancer and it has been hypothesized that kissing is a possible route for oral-oral acquisition of the bacterium.[61] *H. pylori* infection is common in infants in West Africa whose mothers premasticate their child's food.[62] *H. pylori* has been recovered from the saliva of around 10% of patients positive for gastric *H. pylori*, and familial transmission is possible.[14]

Studies on the possible role of oral hygiene and periodontal disease in *H. pylori* infection have shown that the bacterium is detected fairly consistently from the oral cavity and that the chances of *H. pylori* recurrent gastric infection are increased in patients who have oral *H. pylori*.[63]

The fairly high prevalence of *H. pylori* in saliva supports oral-oral transmission,[64] but subsequent studies have suggested that such transmission is not common.[15]

3.1.10 Legionella

Legionella are Gram-negative bacteria transmitted mainly in aerosolized water, and may cause legionellosis—a pneumonia-type illness—particularly in older or immunocompromised people. There are no data to suggest oral-oral transmission of *Legionella* spp. and, as saliva, by virtue of the amylase content inhibits *Legionella pneumophila* growth,[65] salivary transmission is unlikely.

3.1.11 Mycobacteria
3.1.11.1 Mycobacterium tuberculosis

Tuberculosis (TB) is a reemerging infectious disease, affecting about one-third of the global population, and has been increasing in immunocompromised hosts, including older persons, immunosuppressed people and those with HIV/AIDS. Most infections do not have symptoms (latent tuberculosis) and generally affects not only the lungs but also other parts of the body.

The bacilli *M. tuberculosis* and *M. bovis* are the main causes of TB. Person-to-person transmission of TB occurs mainly via the inhalation of droplet nuclei.[66] *Mycobacteria* appear to survive poorly on surfaces over a period of 1 hour.[67]

M. tuberculosis has been identified in saliva from the majority of patients with known TB, and in some with no clinical illness in high-prevalence areas.[17] *M. tuberculosis* has been detected in saliva from patients with TB by direct bacilloscopy, culture, aided by Ziehl-Neelsen or flurochrome staining[68] and nucleic acid amplification.[16,69] *M. bovis*, though classically transmitted in infected cow milk, may also be transmitted by the airborne route.[70] To date, intimate kissing has not been associated with a risk of TB transmission—yet seems to be possible.

3.1.11.2 Atypical Mycobacteria

Atypical mycobacteria (*nontuberculosis mycobacteria*; NTM), also known as environmental mycobacteria, and mycobacteria other than tuberculosis may cause lymphadenitis, though most lymphadenitis is caused by *S. aureus*. Atypical mycobacteriosis also rarely causes bacterial sialadenitis which clinically may resemble a salivary gland tumor.[33] NTM may constitute part of the oral flora of the general population,[71] and infections are increasingly reported in people with HIV/AIDS.[72]

No salivary transmission has been recorded but seems a likely possibility.

3.1.11.3 Mycobacterium leprae

Mycobacterium leprae causes leprosy—a chronic disease characterized by neuropathies, muscle weakness, and disfigurements. About one-third of people with paucibacillary leprosy have *M. leprae* positive in saliva by qPCR,[18] as it had been previously suggested by using conventional PCR.[73] Leprosy is transmitted during close and frequent contacts with untreated cases, transmission being mainly via droplets, usually from the nose but also the mouth, but it is not highly infectious.[74] Most people in leprosy-endemic populations have been exposed to *M. leprae*, yet few develop disease—suggesting that most people develop protective immunity.[75] However, the presence of the *M. leprae* DNA in buccal swabs of leprosy patients and household contacts indicate an additional strategy for dental clinics.[76]

3.1.12 Neisseria

3.1.12.1 Neisseria gonorrhoeae (gonococcus)

Neisseria gonorrhoeae, a Gram-negative coccus, causes gonorrhea, a sexually transmitted infection typified by genital or anorectal infection and discharges. One case report described transmission of gonococcal pharyngitis by intimate kissing.[77] In men who have sex with men who tested positive for pharyngeal gonorrhea, saliva samples were also positive for *N. gonorrhoeae* on culture or nucleic acid amplification tests.[20]

Although we have found no publications that support salivary transmission, this suggested that saliva may transmit gonorrhea without direct throat inoculation in oral-oral, or oro-anogenital intimate practices.[78]

3.1.12.2 Neisseria meningitidis (meningococcus)

Neisseria meningitidis is responsible for some bacterial meningitis—a potentially lethal or life-changing condition. Salivary meningococcal carriage is particularly high among school children in certain areas[79] but a low isolation rate from saliva suggests that low levels of salivary contact (e.g., sharing eating utensils, cups, drinking glasses, or cigarettes) are unlikely to transmit meningococci.[80] However, kissing on the mouth has been implicated as a risk factor in children[81] and intimate kissing has been shown to be a risk factor for the meningococcal carriage found in some university students.[82] A population-based epidemiological study in the United Kingdom on around 14,000 teen-agers found that the higher the number of intimate kissing partners the higher the meningococcal-carrier rates.[19] A cohort study on adolescents (15–19 years of age) with meningococcal disease and matched controls also found that intimate kissing with multiple partners was a significant independent risk factor for meningococcal disease.[83] Intimate kissing appears to increase the risk of *N. meningitidis* transmission and it has therefore been suggested that chemoprophylaxis should be given to household members and anyone kissing and saliva-exchanging contacts of cases of meningococcal meningitis.[84] Finally, and anecdotally some cases of persons who developed a painful eye from meningococcal conjunctivitis 2 days after being spat on in his face further demonstrates the infective potential of saliva.[85,86]

3.1.13 Pasteurella

Human and animal bites most commonly lead to polymicrobial bacterial infections involving a mixture of aerobic and anaerobic organisms. *Staphylococcus, Streptococcus*, and anaerobic bacterial species are common to all mammal bites. *Eikenella* is characteristic of human bite wounds. *Pasteurella* spp. are commonly found in dog and cat bite wounds. *Pasteurella multocida* is a frequent cause of infection after dog or cat bites; most pet cats harboring this bacterial species in their saliva.[87] The risk of salivary transmission from pets to humans seems to be negligible in the absence of bites,[1] but *Pasteurella stomatis* has been isolated from the oral cavity of pet dogs and cats owners.[87] However, human-human transmission is unlikely.

3.1.14 Pseudomonads

Pseudomonads are Gram-negative aerobes often found in biofilms, which can cause a range of issues, from periodontitis and occasional oral infections to pneumonia and septicemia. One outbreak of *P. aeruginosa* infection originated from contaminated mouth swabs probably due to biofilm formation in the wet part of the production line has been described in Norway.[88] *Pseudomonas aeruginosa, Klebsiella pneumoniae, Serratia marcescens, Acinetobacter anitratus*, and *Enterobacter cloacae* have been found in mouthwashes, "Clinifeeds" and residual water from nasogastric aspiration apparatus, probably originating from sinks providing a reservoir of epidemic strains.[89]

P. aeruginosa appears to first establish in the nasopharynx, and its presence in nasal, paranasal sinuses, and oral cavity may predict subsequent bronchopulmonary colonization.[90] Aerobic and anaerobic bacteria can be detected in nearly half of patients with chronic maxillary sinusitis and may include *P. aeruginosa.*[91]

P. aeruginosa is detected in 40% of all oral samples.[5] Oral *pseudomonads*, especially *P. aeruginosa*, may be found particularly when oral hygiene is poor and there is periodontitis, sometimes associated with *Acinetobacteria.*[5] *P. aeruginosa* has been observed subgingivally in patients with chronic or aggressive periodontitis[92] and in the subgingival microbiota of subjects with clinically healthy periodontal tissues especially in HIV/AIDS.[93] *Pseudomonads* as well as a great diversity of other bacteria and yeasts have been isolated in failing dental implants with "peri-implantitis."[94] Bacterial parotitis due to *P. aeruginosa* has been described,[95,96] and neonatal suppurative parotitis is an uncommon disease due to this bacteria.[97] Jaw osteomyelitis generally results from polymicrobial infection, occasionally from *P. aeruginosa.*[98–100] Oral lesions such as carcinoma sites may also harbor *Pseudomonas* spp. together with other bacteria and with yeasts.[57] Radiotherapy to the head and neck region leads to significantly increased of *P. aeruginosa, Streptococci, Candida albicans* on saliva while *Neisseria* and *Actinobacillus* may decrease on the irradiated region.[101] Aerobic and facultative Gram-negative bacilli increased at specific oral sites in patients under myelosuppressive chemotherapy, acute nonlymphocytic leukemia, and small-cell lung carcinoma patients; *P. aeruginosa* and *K. pneumoniae* may be found in some sites, though the most commonly recovery isolates are nonpathogenic *Pseudomonas* spp., especially *Pseudomonas pickettii.*[102]

Microorganisms from dental plaque or periodontal diseases may eventually prove responsible sources for aspiration pneumonia in susceptible individuals.[103] *P. aeruginosa* flourishes in hospital environments, and in hospitalized patients, along with *Staphylococcus* spp. and *Dialister pneumosintes.*[27] Biofilms on endotracheal tubes may lead to ventilator-associated pneumonia. In tubes in place for 23 days, 95% of the sequences belonged to *P. aeruginosa.*[104]

The oral cavity of hospitalized patients may thus be a reservoir for respiratory pathogens which colonize the oropharynx and with alterations in the saliva—especially in prolonged nasogastric tube feeding—

predisposing to aspiration pneumonia.[56] *P. aeruginosa* was cultured from 30% of the subjects on nasogastric tube feeding and 10% of subjects on percutaneous enterogastric tube feeding, but in none of patients fed orally.[105]

The potential respiratory pathogens cultured from medical intensive care unit patients can include *P. aeruginosa*, several genera of Gram-negative bacilli and methicillin-resistant *S. aureus* from the oral cavity.[106] In a pediatric intensive care unit, oral Gram-negative bacteria such as *P. aeruginosa, A. baumannii, K. pneumoniae*, and *Enterobacter* spp. were the predominant pathogens.[107]

P. aeruginosa colonization is especially common in hospitalized patients with cystic fibrosis (CF).[108] Lung destruction—the principal cause of death in CF—may be caused by chronic *P. aeruginosa* infection. The same *P. aeruginosa* clonal types are found in saliva and sputum samples suggesting that the oral cavity is a source for lung infection.[109] The major pathogens are the mucoid variants of *P. aeruginosa*,[110] which have been found in CF from the tongue, buccal mucosa, and saliva.[111,112]

Facial infections or oral colonization with *P. aeruginosa* may rarely lead to septicemia[113,114] especially in immunocompromised people. For example, the oral cavity may be a port of entry of septicemia in patients with hematological malignancies.[115]

Noma neonatorum in premature infants, characterized by gangrene of the nose, mouth, and occasionally the scrotum and eyelids and is typically lethal, is associated with *Pseudomonas* spp. including *P. aeruginosa*,[116–118] as is the noma that may follow cytotoxic chemotherapy.[119] Oral antiseptic decontamination can reduce the salivary and oropharyngeal aerobic flora in artificially ventilated patients, and it decreases but does not reduce multiresistant *Pseudomonas, Acinetobacter*, and *Enterobacter* species.[53] Povidone–iodine disinfection of the upper airway may be effective in reducing *P. aeruginosa* and methicillin-resistant *S. aureus* (MRSA).[120] A combination of chemical oral and mechanical cleansing can significantly reduce potential respiratory pathogens such as methicillin-sensitive *Staphylococcus* spp., *S. pneumoniae*, and *H. influenzae* but, in subjects who underwent systemic antibiotic (Cephazolin) administration without oral cleansing, there was only a nonsignificant reduction of *P. aeruginosa*, methicillin-resistant *Staphylococcus* spp., *S. pneumoniae*, or *H. influenzae*.[121]

Salivary *pseudomonads* are thus ubiquitous particularly in hospitals and hospitalized people, and can spread endogenously to and from the respiratory tracts. Whether they can be transmitted via oral-oral contact is unclear, though likely.

3.1.15 Spirochaetes

Treponema species are not uncommon in the salivary microbiota of HIV/AIDS.[8] *T. parvum* predominates in patients with periodontitis and *T. leuthinolyticum* are more usually isolated in the periodontal health HIV/AIDS group.[8] *T. putidum* are mainly seen in acute necrotic ulcerative gingivitis[122] and *T. denticola* are detected with high prevalence from subgingival biofilms samples in HIV-infected patients with necrotizing periodontal disease.[123]

T. pallidum causes syphilis, an increasingly common sexually transmitted infection,[124,125] potentially lethal from cardiovascular or neurological complications. Syphilis can be transmitted by direct contact with lesions and via mucocutaneous breaks, blood, or saliva.[126] *T. pallidum* is detectable by dark-field microscopy or immunostaining on the surfaces of mouth and other lesions.

The relationship between syphilis and saliva dates to the 18th century, when many hospitals had "salivating rooms" in which patients with syphilis were given mercury in order to induce profuse sweating and salivation to facilitate elimination of the venereal condition.[127] Saliva from patients with syphilis taken directly from the parotid duct failed to show *T. pallidum* which suggests that infectivity was due to the presence of oral lesions.[128] Syphilis can be transmitted by the mouth-to-mouth transfer of prechewed food from actively infected people,[129] and there are case reports in which kissing may have been a route of transmission.[130,131] In a series of 125 children aged 10 years or younger who had acquired syphilis, it was suggested that 23% of cases were due to innocent kissing or to household contact.[132] Acquired syphilis in children, however, is transmitted almost exclusively via sexual abuse, and transmission routes therefore take on particular significance.

Some authors maintain that syphilis can clearly be transmitted by kissing[133] but surprisingly, only isolated case reports have been published[134] and there is as yet no robust evidence of the presence of *T. pallidum* in saliva nor of oral-oral transmission via saliva.

3.1.16 Staphylococcus aureus

Infections caused by *Staphylococcus aureus* range from mild skin infections, to more severe diseases such as osteomyelitis, necrotizing pneumonia, bacteremia leading to endocarditis, septic shock, and septic arthritis.[135]

Oral *S. aureus* is usually isolated in oral mucosa infections, such as angular cheilitis, denture stomatitis, and mucositis, or deep infections, such as osteomyelitis and parotitis.[22,136] An oral carriage rate near to 50% has been reported among healthy students[136] and *S. aureus* salivary load increases in immune-compromised patients.[137]

S. aureus along with other opportunistic pathogens, such as *Pseudomonads, S. pneumoniae, K. pneumoniae*, and *E. coli*, have been detected in the mouths of older patients with pneumonia—particularly when the saliva contains occult blood.[21] The potential respiratory pathogens cultured from medical ICU patients include methicillin-resistant *S. aureus, P. aeruginosa*, and other Gram-negative bacilli.[106] *Staphylococcus* spp. notably are found in saliva of most hospitalized patients, often along with *Pseudomonas* spp., *Acinetobacter* spp., and *D. pneumosintes*.[27] Coagulase-negative *Staphylococci* and *S. aureus* were the most commonly identified bacteria on mobile communication devices used in hospitals and most were methicillin resistant (MRSA)[138] and could be one source. *S. aureus* may also colonize oral lesions such as carcinomas[57] and is the usual cause of bacterial sialadenitis.[33]

Most viridians group *streptococci* have bacteriocin-like activity via hydrogen peroxide and kill MRSA.[139]

There appears to be no evidence of salivary transmission but it may be likely.

3.1.17 Streptococcus

Oral *Streptococcus* spp. can cause severe odontogenic maxillofacial infections[140] and they represent a reservoir for systemic dissemination of pathogenic bacteria and their toxins, leading to infections (e.g., endocarditis) and inflammation (e.g., atherosclerotic disease) in distant body sites.[141]

S. mitis—considered opportunistic infective agents—are common in saliva samples of dentulous adults[142] and may be found in the saliva of hospitalized HIV/AIDS patients.[8]

S. pneumoniae may be detected in saliva from patients with pneumonia;[21] saliva may induce its growth and it typically transmits through droplet spread,[143] supporting evidence that *S. pneumoniae* can use human saliva as a vector for transmission.

S. pyogenes can cause pharyngitis, wound, and other infections. *S. pyogenes* can be carried in saliva by healthy children at school or in day care centers, but rarely so in adults.[23] *S. pyogenes* can particularly be detected in the saliva during acute streptococcal pharyngitis.[22] Intrafamilial spread of group A streptococci has been reported.[144] The transmission of streptococcal pharyngitis to children from adults who prechew the child's food strongly suggests that saliva can act as a vehicle,[145] but kissing has not been proved to be the route of infection.

Nutritionally variant streptococci, Abiotrophia, and *Granulicatella* are implicated in some endocarditis and in bacteremias in cancer patients with neutropenia and mucositis or gingivitis and may be found in saliva.[146]

3.2 CLOSING REMARKS AND PERSPECTIVES

A considerable number of bacteria responsible for medical diseases can be present in saliva and may be transmitted to individuals in close contact with the host. For years, it has been suggested that certain bacterial diseases such as syphilis can be transmitted by kissing, even though there is no robust evidence to support this assertion. We have now accumulated a certain volume of scientific evidence to indicate that meningococcal meningitis can be transmitted by kissing. However, some of these proposals have been based on epidemiological studies in which it was concluded that kissing was a route of transmission of a certain disease based on questionnaires on sexual behavior; the reliability of such questionnaires is limited, especially in the context of infections that can also be transmitted by sexual contact, which can never be fully excluded. Rigorous studies using molecular biology techniques are needed to establish genetic homology between the microorganisms isolated from saliva samples from kissing partners and more definitively establish which diseases can be transmitted by kissing. Only in this way we will be able to design appropriate strategies to control person-to-person infection via this route.

REFERENCES

1. Kennedy SA, Stoll LE, Lauder AS. Human and other mammalian bite injuries of the hand: evaluation and management. *J Am Acad Orthop Surg* 2015;**23**(1):47−57.

2. Bartecchi CE. Cardiopulmonary resuscitation--an element of sophistication in the 18th century. *Am Heart J* 1980;**100**(4):580−1.

3. Arend CF. Transmission of infectious diseases through mouth-to-mouth ventilation: evidence-based or emotion-based medicine?. *Arq Bras Cardiol* 2000;**74**(1):86−97.

4. Edwards S, Carne C. Oral sex and transmission of non-viral STIs. *Sex Transm Infect* 1998; **74**(2):95−100.

5. Souto R, Silva-Boghossian CM, Colombo AP. Prevalence of *Pseudomonas aeruginosa* and *Acinetobacter* spp. in subgingival biofilm and saliva of subjects with chronic periodontal infection. *Braz J Microbiol* 2014;**45**(2):495−501.

6. Mahendran V, Tan YS, Riordan SM, et al. The prevalence and polymorphisms of zonula occludens toxin gene in multiple *Campylobacter concisus* strains isolated from saliva of patients with inflammatory bowel disease and controls. *PLoS One* 2013;**8**(9):e75525.

7. Zhang L, Budiman V, Day AS, et al. Isolation and detection of *Campylobacter concisus* from saliva of healthy individuals and patients with inflammatory bowel disease. *J Clin Microbiol* 2010;**48**(8):2965−7.

8. Zhang F, He S, Jin J, Dong G, Wu H. Exploring salivary microbiota in AIDS patients with different periodontal statuses using 454 GS-FLX titanium pyrosequencing. *Front Cell Infect Microbiol* 2015;**5**:55.

9. Ferraro CT, Gornic C, Barbosa AS, Peixoto RJ, Colombo AP. Detection of *Dialister pneumosintes* in the subgingival biofilm of subjects with periodontal disease. *Anaerobe* 2007; **13**(5-6):244−8.

10. Bajaj JS, Betrapally NS, Hylemon PB, et al. Salivary microbiota reflects changes in gut microbiota in cirrhosis with hepatic encephalopathy. *Hepatology* 2015;**62**(4):1260−71.

11. Reddy S, Sanjai K, Kumaraswamy J, Papaiah L, Jeevan M. Oral carriage of *Enterobacteriaceae* among school children with chronic nail-biting habit. *J Oral Maxillofac Pathol* 2013;**17**(2):163−8.

12. Foweraker JE, Cooke NJ, Hawkey PM. Ecology of *Haemophilus influenzae* and *Haemophilus parainfluenzae* in sputum and saliva and effects of antibiotics on their distribution in patients with lower respiratory tract infections. *Antimicrob Agents Chemother* 1993; **37**(4):804−9.

13. Hasan NA, Young BA, Minard-Smith AT, et al. Microbial community profiling of human saliva using shotgun metagenomic sequencing. *PLoS One* 2014;**9**(5):e97699.

14. Ferguson Jr DA, Li C, Patel NR, Mayberry WR, Chi DS, Thomas E. Isolation of *Helicobacter pylori* from saliva. *J Clin Microbiol* 1993;**31**(10):2802−4.

15. Parsonnet J, Shmuely H, Haggerty T. Fecal and oral shedding of *Helicobacter pylori* from healthy infected adults. *JAMA* 1999;**282**(23):2240−5.

16. Gonzalez Mediero G, Vazquez Gallardo R, Perez Del Molino ML, Diz Dios P. Evaluation of two commercial nucleic acid amplification kits for detecting *Mycobacterium tuberculosis* in saliva samples. *Oral Dis* 2015;**21**(4):451−5.

17. Palakuru SK, Lakshman VK, Bhat KG. Microbiological analysis of oral samples for detection of *Mycobacterium tuberculosis* by nested polymerase chain reaction in tuberculosis patients with periodontitis. *Dent Res J (Isfahan)* 2012;**9**(6):688−93.

18. Rosa FB, Souza VC, Almeida TA, et al. Detection of *Mycobacterium leprae* in saliva and the evaluation of oral sensitivity in patients with leprosy. *Mem Inst Oswaldo Cruz* 2013;**108**(5):572−7.

19. MacLennan J, Kafatos G, Neal K, et al. Social behavior and meningococcal carriage in British teenagers. *Emerg Infect Dis* 2006;**12**(6):950−7.

20. Chow EP, Lee D, Tabrizi SN, et al. Detection of *Neisseria gonorrhoeae* in the pharynx and saliva: implications for gonorrhoea transmission. *Sex Transm Infect* 2016;**92**(5):347−9.

21. Ogawa T, Ikebe K, Enoki K, Murai S, Maeda Y. Investigation of oral opportunistic pathogens in independent living elderly Japanese. *Gerodontology* 2012;**29**(2):e229−33.

22. Dahlén G. Bacterial infections of the oral mucosa. *Periodontol 2000* 2009;**49**(1):13−38.

23. Tagg JR, Ragland NL, Dickson NP. A longitudinal study of lancefield group A streptococcus acquisitions by a group of young Dunedin schoolchildren. *N Z Med J* 1990;**103**(897):429−31.

24. Fenollar F, Lagier JC, Raoult D. *Tropheryma whipplei* and Whipple's disease. *J Infect* 2014;**69**(2):103−12.

25. Dutly F, Hinrikson HP, Seidel T, Morgenegg S, Altwegg M, Bauerfeind P. *Tropheryma whippelii* DNA in saliva of patients without Whipple's disease. *Infection* 2000;**28**(4):219−22.

26. Maragakis LL, Perl TM. *Acinetobacter baumannii*: epidemiology, antimicrobial resistance, and treatment options. *Clin Infect Dis* 2008;**46**(8):1254−63.

27. Zuanazzi D, Souto R, Mattos MB, et al. Prevalence of potential bacterial respiratory pathogens in the oral cavity of hospitalised individuals. *Arch Oral Biol* 2010;**55**(1):21−8.

28. Diaz PI, Hong BY, Frias-Lopez J, et al. Transplantation-associated long-term immunosuppression promotes oral colonization by potentially opportunistic pathogens without impacting other members of the salivary bacteriome. *Clin Vaccine Immunol* 2013;**20**(6):920−30.

29. Sakurai K, Wang D, Suzuki J, et al. High incidence of Actinobacillus actinomycetemcomitans infection in acute coronary syndrome. *Int Heart J* 2007;**48**(6):663−75.

30. Murakami M, Suzuki J, Yamazaki S, et al. High incidence of *Aggregatibacter actinomycetemcomitans* infection in patients with cerebral infarction and diabetic renal failure: a cross-sectional study. *BMC Infect Dis* 2013;**13**:557.

31. Russo T. Agents of actinomycosis. In: Mandel G, Bennett J, Dolin R, editors. *Principles and practices of infectious diseases*. 5th ed. Philadelphia,PA: Churchill Livingston; 2000. p. 2645−54.

32. Hansen T, Kunkel M, Springer E, et al. Actinomycosis of the jaws—histopathological study of 45 patients shows significant involvement in bisphosphonate-associated osteonecrosis and infected osteoradionecrosis. *Virchows Arch* 2007;**451**(6):1009−17.

33. Maier H, Tisch M. Bacterial sialadenitis. *HNO* 2010;**58**(3):229−36.

34. Schneider T, Moos V, Loddenkemper C, Marth T, Fenollar F, Raoult D. Whipple's disease: new aspects of pathogenesis and treatment. *Lancet Infect Dis* 2008;**8**(3):179−90.

35. Street S, Donoghue HD, Neild GH. *Tropheryma whippelii* DNA in saliva of healthy people. *Lancet* 1999;**354**(9185):1178–9.

36. Klotz SA, Ianas V, Elliott SP. Cat-scratch disease. *Am Fam Phys* 2011;**83**(2):152–5.

37. Losch B, Wank R. Life-threatening angioedema of the tongue: the detection of the RNA of *B. henselae* in the saliva of a male patient and his dog as well as of the DNA of three *Bartonella* species in the blood of the patient. *BMJ Case Rep* 2014;**2014** bcr2013203107.

38. Tanner A, Badger S, Lai C, Listgarten M, Visconti R, Socransky S. Wolinella gen. nov., *Wolinella succinogenes* (vibrio succinogenes wolin et al.) comb. nov., and description of *Bacteroides gracilis* sp. nov., *Wolinella recta* sp. nov., *Campylobacter concisus* sp. nov., and *Eikenella corrodens* from humans with periodontal disease. *Int J Syst Evol Microbiol* 1981; **31**(4):432–45.

39. Macfarlane S, Furrie E, Macfarlane GT, Dillon JF. Microbial colonization of the upper gastrointestinal tract in patients with Barrett's esophagus. *Clin Infect Dis* 2007;**45**(1):29–38.

40. Kaakoush NO, Mitchell HM, Man SM. Role of emerging *Campylobacter* species in inflammatory bowel diseases. *Inflamm Bowel Dis* 2014;**20**(11):2189–97.

41. Kaakoush NO, Mitchell HM. *Campylobacter concisus*—a new player in intestinal disease. *Front Cell Infect Microbiol* 2012;**2**:4.

42. Hamasuna R, Hoshina S, Imai H, Jensen JS, Osada Y. Usefulness of oral wash specimens for detecting *Chlamydia trachomatis* from high-risk groups in Japan. *Int J Urol* 2007; **14**(5):473–5.

43. Genc M, Bergman S, Froman G, Elbagir AN, Mardh PA. Antichlamydial activity of saliva. *APMIS* 1990;**98**(5):432–6.

44. Mahmoud EA, Froman G, Genc M, Mardh PA. Age-dependent antichlamydial activity of human saliva: a study of infants, children and adults. *APMIS* 1993;**101**(4):306–10.

45. Moore WE, Holdeman LV, Smibert RM, et al. Bacteriology of experimental gingivitis in children. *Infect Immun* 1984;**46**(1):1–6.

46. Moore W, Moore LV. The bacteria of periodontal diseases. *Periodontol 2000* 1994; **5**(1):66–77.

47. Contreras A, Doan N, Chen C, Rusitanonta T, Flynn MJ, Slots J. Importance of *Dialister pneumosintes* in human periodontitis. *Oral Microbiol Immunol* 2000;**15**(4):269–72.

48. Goldstein EJ, Citron DM, Finegold SM. Role of anaerobic bacteria in bite-wound infections. *Rev Infect Dis* 1984;**6**(Suppl. 1):S177–83.

49. Leao MV, Cassia RC, Santos SS, Silva CR, Jorge AO. Influence of consumption of probiotics on presence of enterobacteria in the oral cavity. *Braz Oral Res* 2011;**25**(5):401–6.

50. Laforce FM, Hopkins J, Trow R, Wang WL. Human oral defenses against Gram-negative rods. *Am Rev Respir Dis* 1976;**114**(5):929–35.

51. Ayars GH, Altman LC, Fretwell MD. Effect of decreased salivation and pH on the adherence of *Klebsiella* species to human buccal epithelial cells. *Infect Immun* 1982;**38**(1):179–82.

52. Kanazuru T, Sato EF, Nagata K, et al. Role of hydrogen generation by *Klebsiella pneumoniae* in the oral cavity. *J Microbiol* 2010;**48**(6):778–83.

53. Fourrier F, Dubois D, Pronnier P, et al. Effect of gingival and dental plaque antiseptic decontamination on nosocomial infections acquired in the intensive care unit: a double-blind placebo-controlled multicenter study. *Crit Care Med* 2005;**33**(8):1728–35.

54. Leao-Vasconcelos LS, Lima AB, Costa Dde M, et al. *Enterobacteriaceae* isolates from the oral cavity of workers in a Brazilian oncology hospital. *Rev Inst Med Trop Sao Paulo* 2015;**57**(2):121–7.

55. Goldberg S, Cardash H, Browning 3rd H, Sahly H, Rosenberg M. Isolation of Enterobacteriaceae from the mouth and potential association with malodor. *J Dent Res* 1997;**76**(11):1770−5.

56. Leibovitz A, Plotnikov G, Habot B, Rosenberg M, Segal R. Pathogenic colonization of oral flora in frail elderly patients fed by nasogastric tube or percutaneous enterogastric tube. *J Gerontol A Biol Sci Med Sci* 2003;**58**(1):52−5.

57. Byakodi R, Krishnappa R, Keluskar V, Bagewadi A, Shetti A. The microbial flora associated with oral carcinomas. *Quintessence Int* 2011;**42**(9):e118−23.

58. Gaetti-Jardim EJ, Ciesielski FI, de Sousa FR, Nwaokorie F, Schweitzer CM, Avila-Campos MJ. Occurrence of yeasts, pseudomonads and enteric bacteria in the oral cavity of patients undergoing head and neck radiotherapy. *Braz J Microbiol* 2011;**42**(3):1047−55.

59. Mackowiak PA, Martin RM, Smith JW. The role of bacterial interference in the increased prevalence of oropharyngeal gram-negative bacilli among alcoholics and diabetics. *Am Rev Respir Dis* 1979;**120**(3):589−93.

60. Kawakami Y, Okimura Y, Kanai M. Prevalence and biochemical properties of Haemophilus species in the oral cavity of healthy adults—investigation of three Japanese individuals. *Microbiol Immunol* 1984;**28**(11):1261−5.

61. Herrera AG. *Helicobacter pylori* and food products: a public health problem. *Methods Mol Biol* 2004;**268**:297−301.

62. Megraud F. Transmission of *Helicobacter pylori*: faecal-oral versus oral-oral route. *Aliment Pharmacol Ther* 1995;**9**(Suppl. 2):85−91.

63. Anand PS, Kamath KP, Anil S. Role of dental plaque, saliva and periodontal disease in *Helicobacter pylori* infection. *World J Gastroenterol* 2014;**20**(19):5639−53.

64. Li C, Ha T, Ferguson DA, et al. A newly developed PCR assay of *H. pylori* in gastric biopsy, saliva, and feces. *Dig Dis Sci* 1996;**41**(11):2142−9.

65. Bortner CA, Miller RD, Arnold RR. Effects of alpha-amylase on in vitro growth of *Legionella pneumophila*. *Infect Immun* 1983;**41**(1):44−9.

66. Cruz-Knight W, Blake-Gumbs L. Tuberculosis: an overview. *Prim Care* 2013;**40**(3):743−56.

67. Lever MS, Williams A, Bennett AM. Survival of mycobacterial species in aerosols generated from artificial saliva. *Lett Appl Microbiol* 2000;**31**(3):238−41.

68. Holani AG, Ganvir SM, Shah NN, et al. Demonstration of *Mycobacterium tuberculosis* in sputum and saliva smears of tuberculosis patients using Ziehl−Neelsen and flurochrome staining—a comparative study. *J Clin Diagn Res* 2014;**8**(7):ZC42−5.

69. Eguchi J, Ishihara K, Watanabe A, Fukumoto Y, Okuda K. PCR method is essential for detecting *Mycobacterium tuberculosis* in oral cavity samples. *Oral Microbiol Immunol* 2003;**18**(3):156−9.

70. Evans JT, Smith EG, Banerjee A, et al. Cluster of human tuberculosis caused by *Mycobacterium bovis*: evidence for person-to-person transmission in the UK. *Lancet* 2007;**369**(9569):1270−6.

71. Mills CC. Occurrence of mycobacterium other than *Mycobacterium tuberculosis* in the oral cavity and in sputum. *Appl Microbiol* 1972;**24**(3):307−10.

72. Elvira J, Garcia del Rio E, Lopez-Suarez A, Garcia-Martos P, Giron JA. Submandibular gland infection by *Mycobacterium avium-intracellulare* in an AIDS patient. *Eur J Clin Microbiol Infect Dis* 1998;**17**(7):529−35.

73. Abdalla LF, Santos JHA, Collado CSC, Cunha MGS, Naveca FG. *Mycobacterium leprae* in the periodontium, saliva and skin smears of leprosy patients. *Rev Odonto Ciênc* 2010;**25**(2):148−53.

74. Smith WC, Smith CM, Cree IA, et al. An approach to understanding the transmission of *Mycobacterium leprae* using molecular and immunological methods: results from the MILEP2 study. *Int J Lepr Other Mycobact Dis* 2004;**72**(3):269–77.

75. Ramaprasad P, Fernando A, Madhale S, et al. Transmission and protection in leprosy: Indications of the role of mucosal immunity. *Lepr Rev* 1997;**68**(4):301–15.

76. Martinez TS, Figueira MM, Costa AV, Goncalves MA, Goulart LR, Goulart IM. Oral mucosa as a source of *Mycobacterium leprae* infection and transmission, and implications of bacterial DNA detection and the immunological status. *Clin Microbiol Infect* 2011; **17**(11):1653–8.

77. Willmott FE. Transfer of gonococcal pharyngitis by kissing? *Br J Vener Dis* 1974; **50**(4):317–18.

78. Chow EP, Tabrizi SN, Phillips S, et al. Neisseria gonorrhoeae bacterial DNA load in the pharynges and saliva of men who have sex with men. *J Clin Microbiol* 2016;**54**(10):2485–90.

79. Martinez I, Lopez O, Sotolongo F, Mirabal M, Bencomo A. Carriers of *Neisseria meningitidis* among children from a primary school. *Rev Cubana Med Trop* 2003;**55**(3):162–8.

80. Orr HJ, Gray SJ, Macdonald M, Stuart JM. Saliva and meningococcal transmission. *Emerg Infect Dis* 2003;**9**(10):1314–15.

81. Stanwell-Smith RE, Stuart JM, Hughes AO, Robinson P, Griffin MB, Cartwright K. Smoking, the environment and meningococcal disease: a case control study. *Epidemiol Infect* 1994;**112**(2):315–28.

82. Neal KR, Nguyen-Van-Tam JS, Jeffrey N, et al. Changing carriage rate of *Neisseria meningitidis* among university students during the first week of term: cross sectional study. *BMJ* 2000;**320**(7238):846–9.

83. Tully J, Viner RM, Coen PG, et al. Risk and protective factors for meningococcal disease in adolescents: matched cohort study. *BMJ* 2006;**332**(7539):445–50.

84. Cuevas LE, Hart CA. Chemoprophylaxis of bacterial meningitis. *J Antimicrob Chemother* 1993;**31**(Suppl. B):79–91.

85. Holdsworth G, Jackson H, Kaczmarski E. Meningococcal infection from saliva. *Lancet* 1996;**348**(9039):1443.

86. Dryden AW, Rana M, Pandey P. Primary meningococcal conjunctivitis: an unusual case of transmission by saliva. *Digit J Ophthalmol* 2016;**22**(1):25–7.

87. Arashima Y, Kumasaka K, Okuyama K, et al. Clinicobacteriological study of *Pasteurella multocida* as a zoonosis (1). Condition of dog and cat carriers of *Pasteurella*, and the influence for human carrier rate by kiss with the pets. *Kansenshogaku Zasshi* 1992;**66**(2):221–4.

88. Iversen BG, Eriksen HM, Bo G, et al. *Pseudomonas aeruginosa* contamination of mouth swabs during production causing a major outbreak. *Ann Clin Microbiol Antimicrob* 2007;**6**:3.

89. Millership SE, Patel N, Chattopadhyay B. The colonization of patients in an intensive treatment unit with gram-negative flora: the significance of the oral route. *J Hosp Infect* 1986; **7**(3):226–35.

90. Rivas Caldas R, Boisrame S. Upper aero-digestive contamination by *Pseudomonas aeruginosa* and implications in cystic fibrosis. *J Cyst Fibros* 2015;**14**(1):6–15.

91. Paju S, Bernstein JM, Haase EM, Scannapieco FA. Molecular analysis of bacterial flora associated with chronically inflamed maxillary sinuses. *J Med Microbiol* 2003; **52**(Pt 7):591–7.

92. Silva-Boghossian CM, Neves AB, Resende FA, Colombo AP. Suppuration-associated bacteria in patients with chronic and aggressive periodontitis. *J Periodontol* 2013;**84**(9):e9–e16.

93. Goncalves LS, Souto R, Colombo AP. Detection of *Helicobacter pylori*, *Enterococcus faecalis*, and *Pseudomonas aeruginosa* in the subgingival biofilm of HIV-infected subjects undergoing HAART with chronic periodontitis. *Eur J Clin Microbiol Infect Dis* 2009; **28**(11):1335–42.

94. Alcoforado GA, Rams TE, Feik D, Slots J. Microbial aspects of failing osseointegrated dental implants in humans. *J Parodontol* 1991;**10**(1):11–18.

95. Pruett TL, Simmons RL. Nosocomial gram-negative bacillary parotitis. *JAMA* 1984; **251**(2):252–3.

96. Vassal O, Bernet C, Wallet F, Friggeri A, Piriou V. Bacterial parotitis in an immunocompromised patient in adult ICU. *Ann Fr Anesth Reanim* 2013;**32**(9):615–17.

97. Ozdemir H, Karbuz A, Ciftci E, Fitoz S, Ince E, Dogru U. Acute neonatal suppurative parotitis: a case report and review of the literature. *Int J Infect Dis* 2011;**15**(7):e500–2.

98. Hoen MM, Downs RH, LaBounty GL, Nespeca JA. Osteomyelitis of the maxilla with associated vertical root fracture and pseudomonas infection. *Oral Surg Oral Med Oral Pathol* 1988;**66**(4):494–8.

99. Pappalardo S, Tanteri L, Brutto D, et al. Mandibular osteomyelitis due to *Pseudomonas aeruginosa*. Case report. *Minerva Stomatol* 2008;**57**(6):323–9.

100. Persac S, Peron JM. Temporal cellulitis of odontogenic origin complicated by temporomandibular osteoarthritis. *Rev Stomatol Chir Maxillofac* 2008;**109**(2):110–13.

101. Liu K, Gao N, Wang YC, et al. The changes of bacteria group on oral mucosa after radiotherapy of postoperative patients of oral carcinoma. *Hua Xi Kou Qiang Yi Xue Za Zhi* 2005;**23**(2):128–9.

102. Minah GE, Rednor JL, Peterson DE, Overholser CD, Depaola LG, Suzuki JB. Oral succession of Gram-negative bacilli in myelosuppressed cancer patients. *J Clin Microbiol* 1986; **24**(2):210–13.

103. Imsand M, Janssens JP, Auckenthaler R, Mojon P, Budtz-Jorgensen E. Bronchopneumonia and oral health in hospitalized older patients. A pilot study. *Gerodontology* 2002; **19**(2):66–72.

104. Perkins SD, Woeltje KF, Angenent LT. Endotracheal tube biofilm inoculation of oral flora and subsequent colonization of opportunistic pathogens. *Int J Med Microbiol* 2010; **300**(7):503–11.

105. Leibovitz A, Plotnikov G, Habot B, et al. Saliva secretion and oral flora in prolonged nasogastric tube-fed elderly patients. *Isr Med Assoc J* 2003;**5**(5):329–32.

106. Scannapieco FA, Stewart EM, Mylotte JM. Colonization of dental plaque by respiratory pathogens in medical intensive care patients. *Crit Care Med* 1992;**20**(6):740–5.

107. Pedreira ML, Kusahara DM, de Carvalho WB, Nunez SC, Peterlini MA. Oral care interventions and oropharyngeal colonization in children receiving mechanical ventilation. *Am J Crit Care* 2009;**18**(4):319–28.

108. Folkesson A, Jelsbak L, Yang L, et al. Adaptation of *Pseudomonas aeruginosa* to the cystic fibrosis airway: an evolutionary perspective. *Nat Rev Microbiol* 2012;**10**(12):841–51.

109. Rivas Caldas R, Le Gall F, Revert K, et al. *Pseudomonas aeruginosa* and periodontal pathogens in the oral cavity and lungs of cystic fibrosis patients: a case-control study. *J Clin Microbiol* 2015;**53**(6):1898–907.

110. Komiyama K, Tynan JJ, Habbick BF, Duncan DE, Liepert DJ. *Pseudomonas aeruginosa* in the oral cavity and sputum of patients with cystic fibrosis. *Oral Surg Oral Med Oral Pathol* 1985;**59**(6):590–4.

111. Lindemann RA, Newman MG, Kaufman AK. Mucoid variant *Pseudomonas aeruginosa* isolated in the oral cavity of a cystic fibrosis patient. *Spec Care Dentist* 1983;**3**(5):222–3.

112. Lindemann RA, Newman MG, Kaufman AK, Le TV. Oral colonization and susceptibility testing of *Pseudomonas aeruginosa* oral isolates from cystic fibrosis patients. *J Dent Res* 1985;**64**(1):54–7.

113. Weinbren MJ, Forgeson G, Helenglass G, Jameson B, Powles R. Unusual presentation of pseudomonas infection. *BMJ* 1988;**297**(6655):1034–5.

114. Gosney MA, Preston AJ, Corkhill J, Millns B, Martin MV. *Pseudomonas aeruginosa* septicaemia from an oral source. *Br Dent J* 1999;**187**(12):639–40.

115. Bergmann OJ, Kilian M, Ellegaard J. Potentially pathogenic microorganisms in the oral cavity during febrile episodes in immunocompromised patients with haematologic malignancies. *Scand J Infect Dis* 1989;**21**(1):43–51.

116. Ghosal SP, Sen Gupta PC, Mukherjee AK, Choudhury M, Dutta N, Sarkar AK. Noma neonatorum: its aetiopathogenesis. *Lancet* 1978;**2**(8084):289–91.

117. Juster-Reicher A, Mogilner BM, Levi G, Flidel O, Amitai M. Neonatal noma. *Am J Perinatol* 1993;**10**(6):409–11.

118. Freeman AF, Mancini AJ, Yogev R. Is noma neonatorum a presentation of ecthyma gangrenosum in the newborn? *Pediatr Infect Dis J* 2002;**21**(1):83–5.

119. Brady-West DC, Richards L, Thame J, Moosdeen F, Nicholson A. Cancrum oris (noma) in a patient with acute lymphoblastic leukaemia. A complication of chemotherapy induced neutropenia. *West Indian Med J* 1998;**47**(1):33–4.

120. Masaki H, Nagatake T, Asoh N, et al. Significant reduction of nosocomial pneumonia after introduction of disinfection of upper airways using povidone-iodine in geriatric wards. *Dermatology* 2006;**212**(Suppl. 1):98–102.

121. Okuda M, Kaneko Y, Ichinohe T, Ishihara K, Okuda K. Reduction of potential respiratory pathogens by oral hygienic treatment in patients undergoing endotracheal anesthesia. *J Anesth* 2003;**17**(2):84–91.

122. Wyss C, Moter A, Choi BK, et al. *Treponema putidum* sp. nov., a medium-sized proteolytic spirochaete isolated from lesions of human periodontitis and acute necrotizing ulcerative gingivitis. *Int J Syst Evol Microbiol* 2004;**54**(Pt 4):1117–22.

123. Ramos MP, Ferreira SM, Silva-Boghossian CM, et al. Necrotizing periodontal diseases in HIV-infected brazilian patients: a clinical and microbiologic descriptive study. *Quintessence Int* 2012;**43**(1):71–82.

124. Leuci S, Martina S, Adamo D, et al. Oral syphilis: a retrospective analysis of 12 cases and a review of the literature. *Oral Dis* 2013;**19**(8):738–46.

125. Scully C, Setterfield JF. Return of the great pox. *Dent Update* 2016;**43**(3):267–8.

126. Little JW. Syphilis: an update. *Oral Surg Oral Med Oral Pathol Oral Radiol Endod* 2005;**100**(1):3–9.

127. Pandya SK. Salivating rooms. *Natl Med J India* 2005;**18**(5):263–4.

128. Barnett CW, Kulchar GV. The infectivity of saliva in early syphilis. *J Invest Dermatol* 1939;**2**(6):327–9.

129. Zhou P, Qian Y, Lu H, Guan Z. Nonvenereal transmission of syphilis in infancy by mouth-to-mouth transfer of prechewed food. *Sex Transm Dis* 2009;**36**(4):216–17.

130. Brumfield WA, Smith DC. Transmission sequence of syphilis. *Am J Public Health Nations Health* 1934;**24**(6 Pt 1):576–80.

131. Murrell M, Gray M. Acquired syphilis in children. *Br Med J* 1947;**2**(4518):206-207.

132. Smith Jr F. Acquired syphilis in children. An epidemiologic and clinical study. *Am J Syphilis* 1939;**23**:165–85.

133. Tramont E. Treponema pallidum (syphilis). In: Mandell G, Bennett J, Dolin R, editors. *Principles and Practice of Infectious Diseases*. 6th ed. New York, NY: Churchill Livingston; 2005. p. 2768–85.

134. Yu X, Zheng H. Syphilitic chancre of the lips transmitted by kissing: a case report and review of the literature. *Medicine* 2016;**95**(14):e3303.

135. Crossley KB, Jefferson KK, Archer GL, Fowler VG. *Staphylococci in human disease*. 2nd ed. Oxford, UK: Wiley-Blackwell; 2010.

136. Smith AJ, Jackson MS, Bagg J. The ecology of *Staphylococcus* species in the oral cavity. *J Med Microbiol* 2001;**50**(11):940–6.

137. Arirachakaran P, Poovorawan Y, Dahlen G. Highly-active antiretroviral therapy and oral opportunistic microorganisms in HIV-positive individuals of Thailand. *J Investig Clin Dent* 2016;**7**(2):158–67.

138. Morvai J, Szabo R. The role of mobile communication devices in the spread of infections. *Orv Hetil* 2015;**156**(20):802–7.

139. Uehara Y, Kikuchi K, Nakamura T, et al. H(2)O(2) produced by viridans group strepto-cocci may contribute to inhibition of methicillin-resistant *Staphylococcus aureus* colonization of oral cavities in newborns. *Clin Infect Dis* 2001;**32**(10):1408–13.

140. Rasteniene R, Puriene A, Aleksejuniene J, Peciuliene V, Zaleckas L. Odontogenic maxillo-facial infections: a ten-year retrospective analysis. *Surg Infect (Larchmt)* 2015;**16**(3):305–12.

141. Han YW, Wang X. Mobile microbiome: oral bacteria in extra-oral infections and inflam-mation. *J Dent Res* 2013;**92**(6):485–91.

142. Ealla KK, Ghanta SB, Motupalli NK, Bembalgi M, Madineni PK, Raju PK. Comparative analysis of colony counts of different species of oral streptococci in saliva of dentulous, edentulous and in those wearing partial and complete dentures. *J Contemp Dent Pract* 2013;**14**(4):601–4.

143. Verhagen LM, de Jonge MI, Burghout P, et al. Genome-wide identification of genes essen-tial for the survival of *Streptococcus pneumoniae* in human saliva. *PLoS One* 2014;**9**(2):e89541.

144. Tanz RR, Shulman ST. Chronic pharyngeal carriage of group A streptococci. *Pediatr Infect Dis J* 2007;**26**(2):175–6.

145. Steinkuller JS, Chan K, Rinehouse SE. Prechewing of food by adults and streptococcal pharyngitis in infants. *J Pediatr* 1992;**120**(4 Pt 1):563–4.

146. Yacoub AT, Krishnan J, Acevedo IM, Halliday J, Greene JN. Nutritionally variant strepto-cocci bacteremia in cancer patients: a retrospective study, 1999-2014. *Mediterr J Hematol Infect Dis* 2015;**7**(1):e2015030.

CHAPTER 4

Viral Diseases Transmissible by Kissing

Viral infections can be transmitted by various routes. At one extreme, airborne or droplet viral infections (e.g., varicella zoster, ebola) are highly contagious. Most viruses can be spread by touching surfaces contaminated by the virus and then touching the mouth or eyes. Mass gatherings,[1] clinical and chronic care facilities may be hotspots for virus spread when transmission is via aerosols, droplets, or fomites (contaminated surfaces). Environmental factors which are often important for virus survival may include the ambient humidity, temperature, and pH of the environment they are in, so many viruses survive only a few hours in the environment and are often readily inactivated by common hygiene techniques, using soap and water, and some detergents, disinfectants, and antiseptics. Sexually transmitted viral infections, such as herpes simplex, are often transmitted by close mucosal contacts.

Virus infections may be seen especially in immunocompromised people and can be life-changing or even lethal. Some viruses, such as herpes simplex viruses, can also cause obvious oral lesions. Some mainly affect animals but can transmit to man (zoonoses), and viruses may mutate.

4.1 VIRUS IN SALIVA

Various viruses habitually colonize the human mouth and may be present in saliva in quantities sufficient to infect other individuals.[2] Even in instances where ductal saliva might contain no bugs, whole saliva may contain infectious agents from other sites. For example, people who carry respiratory viruses often also have the virus in their saliva[3] as may some who have viral gastroenteritis, and where there is oral bleeding, blood-borne viruses may be present. Humans can be reservoirs of viruses; asymptomatic shedding before clinical disease or where the infection is subclinical (undiagnosed or symptomless) is a major factor in their spread. Host innate or acquired immunity, and saliva can be protective against many infections which may be increased in people

Saliva Protection and Transmissible Diseases. DOI: http://dx.doi.org/10.1016/B978-0-12-813681-2.00004-4
© 2017 Elsevier Inc. All rights reserved.

with immunocompromising states or hyposalivation. Genetic factors play a role in transmission; e.g., the Major Histocompatibilty Complex has a role in *Human Immunodeficiency Virus* (HIV) control, along with virus survival and other factors.

Kissing may spread viral agents from person-to-person if the recipient is susceptible. Kissing often occurs alone while, in sexual interplay, there is usually also kissing, especially deep kissing, and there may be oro-anogenital contact and transmission of viruses.[4]

Viruses responsible for diseases such as hepatitis viruses, herpesvirus infections (e.g., with *Herpes simplex types 1 and 2, Epstein-Barr virus, Cytomegalovirus*, and *Kaposi syndrome herpesvirus*), and *papillomaviruses* can be conveyed by kissing—as can potentially other viruses present in saliva such as *Ebola* and *Zika viruses*.

It may be difficult or impossible to differentiate saliva transmission from that by body fluids or fomites introduced into the mouth, respiratory droplets, aerosols, or other routes. Reliable studies are needed, designed specifically to clarify which diseases can be transmitted in humans by saliva in order to develop appropriate strategies to control person-to-person spread of infection by this route.

In an ideal world, the implication of viruses in diseases is best achieved using a modification of Koch's postulates[5] as follows:

1. A nucleic acid sequence belonging to a putative pathogen should be present in most cases of an infectious disease. Microbial nucleic acids should be found preferentially in those organs or gross anatomic sites known to be diseased, and not in those organs that lack pathology.
2. Fewer, or no, copy numbers of pathogen-associated nucleic acid sequences should occur in hosts or tissues without disease.
3. With resolution of disease, the copy number of pathogen-associated nucleic acid sequences should decrease or become undetectable. With clinical relapse, the opposite should occur.
4. When sequence detection predates disease, or sequence copy number correlates with severity of disease or pathology, the sequence–disease association is more likely to be a causal relationship.
5. The nature of the microorganism inferred from the available sequence should be consistent with the known biological characteristics of that group of organisms.
6. Tissue-sequence correlates should be sought at the cellular level: efforts should be made to demonstrate specific *in situ* hybridization of

microbial sequence to areas of tissue pathology and to visible infection or to areas where microorganisms are presumed to be located.

7. These sequence-based forms of evidence for microbial causation should be reproducible.

Such data for viruses in saliva are sparse so in the meantime epidemiological data and the presence of virus in saliva, nucleic acid, or antigens of microorganism have to be relied upon as circumstantial evidence. Most evidence for the presence of viruses in saliva comes from salivary culture, or serological responses, but nucleic acid amplification techniques such as Reverse Transcription Polymerase Chain Reaction (RT-PCR) can now be used and have detected many previously undetectable virus infections.

Some of the implications of viruses in oral healthcare are reviewed elsewhere[2,6] and antiviral activities of saliva have been reviewed elsewhere[7,8] but this is an exploding field[9] so here we review viruses classification (Table 4.1) and the main human viruses detectable in saliva (Table 4.2).

4.1.1 Adenoviruses (HAdV)

At least 69 HAdV genotypes are recognized. HAdV are a common cause mainly of respiratory infections and, typically in men who have sex with men, a cause of urethritis and conjunctivitis.[50] Respiratory infection caused by HAdV in immunocompetent people is typically caused by HAdV-3, mild and self-limited. However, more recently HAdV-55 in particular has been found to cause severe community-acquired pneumonia and acute respiratory distress syndrome in immunocompetent adults,[51] mainly from China. HAdV pneumonia typically is found in neonates, immunocompromised people, and school or military camp populations. HAdV infections in immunocompromised individuals can be severe and life-threatening. The past decade has also seen the emergence in resource-rich countries of several other new viruses—often in traveling people—including HAdV-14p1.[52] HAdV can be found in saliva[3] and it is presumed that it can transmit infection though the main routes are through respiratory droplets or touching infected objects. *Adenoviruses* can survive a long time on objects and spread easily. Adenoviral shedding in saliva and feces has also been reported after p53 adenoviral gene therapy in some patients with esophageal cancer.[53]

Table 4.1 Virus Classification

Family	Genus
DNA Viruses	
Adenoviridae	*Adenovirus*
Papovaviridae	*Papillomavirus*
Parvoviridae	*B19 parvovirus*
Herpesviridae	*Herpes simplex virus, Varicella zoster virus, Cytomegalovirus, Epstein-Barr virus,* HHV-6, HHV-7, HHV-8
Hepadnaviridae	*Hepatitis B virus*
Polyomaviridae	*Polyomavirus* (progressive multifocal leucoencephalopathy)
RNA Viruses	
Bunyaviridae	*Hantaviruses*
Caliciviridae	*Norwalk viruses, Hepatitis E virus*
Coronaviridae	*Coronaviruses*
Flaviviridae	*Dengue virus, Hepatitis C virus, Yellow fever virus, Zika virus*
Filoviridae	*Ebola virus, Marburg virus*
Orthomyxoviridae	*Influenza virus*
Paramyxoviridae	*Measles virus, Mumps virus, Respiratory syncytial virus*
Picornaviridae	*Poliovirus, Rhinoviruses, Hepatitis A virus*
Reoviridae	*Reovirus, Rotavirus*
Retroviridae	HIV-1, HIV-2, HTLV-I
Rhabdoviridae	*Rabies virus*
Togaviridae	*Rubella virus*

Table 4.2 Medical Viruses Detectable in Saliva (Alphabetical Order)

Virus	Healthy/Asymptomatic Population (Reference)	Active Infection Patients (Reference)
Adenovirus	NAI	Robinson et al. (2008)[3]
Chikungunya	NAI	Gardner et al. (2015)[10]
Coronavirus-MERS	NAI	Goh et al. (2013)[11]
Coronavirus-SARS	NAI	Wang et al. (2004)[12]
Cytomegalovirus (CMV; HHV-5)	Cannon et al. (2014)[13]	Pinninti et al. (2015)[14]
Dengue virus	NAI	Andries et al. (2015)[15]
Ebola virus (EBOV)	NAI	Bausch et al. (2007)[16]
Enteroviruses	Graves et al. (2003)[17]	NAI
Epstein-Barr virus (EBV; HHV-4)	Ikuta et al. (2000)[18]	Balfour et al. (2013)[19]
Hantavirus	NAI	Pettersson et al. (2008)[20]
Hepatitis A virus	NAI	Joshi et al. (2014)[21]
Hepatitis B virus	NAI	Arora et al. (2012)[22]

(Continued)

Table 4.2 (Continued)

Virus	Healthy/Asymptomatic Population (Reference)	Active Infection Patients (Reference)
Hepatitis C virus	NAI	Hermida et al. (2002)[23]
Hepatitis G virus	Yan et al. (2002)[24]	Seemayer et al. (1998)[25]
Herpes simplex virus type 1 (HSV-1; HHV-1)	Miller and Danaher (2008)[26]	Gilbert (2006)[27]
Herpes simplex virus type 2 (HSV-2; HHV-2)	Tateishi et al. (1994)[28]	NAI
Human herpesvirus 6	Zerr et al. (2005)[29]	Leibovitch et al. (2014)[30]
Human herpesvirus 7	Magalhães et al. (2010)[31]	Watanabe et al. (2002)[32]
Human herpesvirus 8 (HHV-8; KSHV)	De Souza et al. (2007)[33]	Vieira et al. (1997)[34]
Human immunodeficiency virus (HIV)	NAI	Navazesh et al. (2010)[35]
Human papillomavirus (HPV)	Kreimer et al. (2010)[36]	Lopez-Villanueva et al. (2011)[37]
Influenza viruses	NAI	Bilder et al. (2011)[38]
Measles virus	NAI	Oliveira et al. (2003)[39]
Metapneumovirus	NAI	NAI
Molluscum contagiosum virus	NAI	NAi
Mumps virus	NAI	Royuela et al. (2011)[40]
Nipah virus	NAI	Luby et al. (2009)[41]
Norovirus	NAI	NAI
Parainfluenza viruses	NAI	NAI
Parvovirus	NAI	NAI
Polyomavirus	Robaina et al. (2013)[42]	Loyo et al. (2010)[43]
Rabies virus	NAI	Crepin et al. (1998)[44]
Respiratory Syncytial Virus (RSV)	NAI	Robinson et al. (2008)[3]
Rhinoviruses	NAI	NAI
Rotaviruses	NAI	NAI
Rubella virus	NAI	Jin et al. (2002)[45]
Torque teno virus	Naganuma et al. (2008)[46]	NAI
Varicella zoster virus (VZV; HHV-3)	Mehta et al. (2004)[47]	Mehta et al. (2008)[48]
West Nile virus	NAI	NAI
Yellow fever virus	NAI	NAI
Zika virus (ZIKV)	NAI	Musso et al. (2015)[49]

NAI, no available information; MERS, Middle East Respiratory Syndrome; SARS, Severe Acute Respiratory Syndrome; KSHV, Kaposi's sarcoma-associated herpesvirus.

4.1.2 Chikungunya Virus (CHIKV)

CHIKV is a *lavivirus* transmitted mainly by *Aedes aegypti* and *Aedes albopictus* mosquitoes, causing fever and joint pains—similar to Dengue fever.

CHIKV can be present in saliva—confirmed by RT-PCR—and especially associated with oral/nasal hemorrhagic lesions in the viremic period.[10] Since more than 50% of CHIKV-infected people experience gingival bleeding,[54] this could also encourage infection transmission by kissing. Nonvector-based mother-to-child transmission of CHIKV has been reported.[55] The impact of potential CHIKV transmission via saliva needs to be more seriously assessed as it could be highly relevant, especially in immunocompromised patients.[56]

4.1.3 Coronaviruses (CoV)

Coronaviruses are common viruses that can infect humans, and animals as diverse as bats and alpacas. There are a number of *Human coronaviruses* and they usually cause respiratory infections—mostly mild illnesses such as the common cold. However, several *coronaviruses* including the Middle East Respiratory Syndrome (MERS), especially seen in Saudi Arabia or visitors to that area, and *Severe Acute Respiratory Syndrome* (SARS), seen mainly in China and travelers from there, can cause more severe and sometimes life-threatening human infections.[52,57] *Coronaviruses* that cause severe acute respiratory infections have >50% mortality rates in older and immunosuppressed people.[58] WIV1-CoV, a virus similar to SARS, could also be poised to cause epidemics.[59]

People living with or caring for someone with a *coronavirus* infection are most at risk of developing the infection themselves. *Coronavirus* transmission is mainly oral–fecal and respiratory from small droplets of saliva or on fomites. Oral–urine and saliva transmission of MERS-CoV and SARS-CoV are also highly likely.[11,12] Salivary cystatin D, a cysteine protease inhibitor, can inhibit replication of some *coronaviruses*.[60] Although evidence is sparse, SARS-CoV appears to be transmitted primarily through saliva droplets. Kissing could constitute a route for transmission.

4.1.4 Dengue Virus (DENV)

DENV, transmitted by *Aedes aegypti* and *Aedes albopictus* mosquitoes, is the most important arthropod-borne flavivirus virus affecting humans and, although most infections are asymptomatic or cause only a mild fever, is capable of producing life-threatening hemorrhagic fever, shock syndrome, and systemic complications (e.g., encephalitis and hepatitis).

DENV has been isolated from human saliva and urine.[15,61−63]

Human salivary transmission of dengue appears to be most unlikely.

4.1.5 Ebola Virus (EBOV)

EBOV is a highly lethal *flavivirus* infection, transmitted via bats and mammalian (monkey and ape) bush meat, causing hemorrhagic fever in humans. EBOV is shed in a wide variety of bodily fluids, including saliva, especially during the acute illness.[16] In a series of eight seriously ill people with Ebola disease, all oral fluid samples obtained 5−10 days after the onset of symptoms were positive for EBOV by RT-PCR.[64] Patients with detectable EBOV in saliva show a higher mortality which likely reflects increased virus shedding in patients with high viremia, an indicator of a poor prognosis.[65]

Human-to-human transmission of EBOV is mainly through direct contact with the tissues, blood, secretions, or other body fluids, including saliva, of infected hosts.[66] Particular concern is the frequent presence of EBOV in saliva early in the course of Ebola disease.[16] Moreover, the person-to-person transmission risk increases—bearing in mind that 1−6% of infected individuals are asymptomatic or mildly symptomatic and the incubation period could last up to 21 days.[66] EBOV could be transmitted to others via saliva by sharing food and through intimate contact, although up to date the only documented cases of secondary transmission from recovered patients have been through sexual transmission.[67] No cases of Ebola transmission through deep kissing have been confirmed, although there must surely be a very high potential risk. In West Africa, the kissing of dead bodies is a traditional burial practice and can promote EBOV transmission.[68]

4.1.6 Enteroviruses (EV)

At least 70 serotypes of EV that can infect humans have been identified, including mainly, *Coxsackieviruses* (*groups A* and *B*), *ECHOviruses*, and *polioviruses*. Recently, the EV genus has been reclassificated in five species: *Human enteroviruses A, B, C,* and *D* and *Poliovirus*. Enteroviruses *a*re implicated in a range of diseases, some of which may affect the mouth, including herpangina and hand-foot and mouth disease (HFMD) which are common and in which complications including pneumonia, meningitis, or encephalitis are seen but rarely. Animals may be affected by similar conditions (such as foot and mouth disease) but these only rarely transmit to man.[69] *Enteroviruses* have also been detected by PCR in saliva of asymptomatic children.[17]

4.1.6.1 Herpangina

Herpangina is typically caused by *Coxsackieviruses* A1 to 6, 8, 10, and 22. Other cases are caused by *Coxsackie* group B (strains 1–4), ECHOviruses, and other *enteroviruses*. Papulovesicular oropharyngeal lesions progress to ulcers and there is usually no rash.[70]

Children with *Coxsackie* A2 infections mostly present with herpangina only, and have fewer central nervous system complications and a better outcome than those with *Enterovirus* 71 (EV71) infections.[71]

4.1.6.2 Hand, Foot, and Mouth Disease

Hand, Foot, and Mouth Disease (HFMD) is an exanthem on the hands and feet with associated fever and oral lesions.[70] HFMD, is typically a mild illness, caused mainly by Coxsackievirus A16 or EV71,[72] occasionally by *Coxsackieviruses A4–7, A9, A10, B1–B3*, or *B5*. Indeed, over 100 serotypes of enterovirus species may cause HFMD.[73]

4.1.6.3 Paraechoviruses (HPeV)

HPeV types 1 and 2, were previously known as *ECHOviruses 22* and *23*, respectively, and can be associated with gastrointestinal, respiratory, or meningeal infections.[74]

4.1.6.4 Poliovirus

Poliomyelitis, which is potentially lethal, is now extremely rare. *Poliovirus* was found in secretions from the upper respiratory tract and salivary swabs of household contacts of patients with virologically proven poliomyelitis.[75]

Enteroviruses are highly contagious, spread mainly by oral–oral and fecal–oral routes, and typically affect children under 10 years old. It has been speculated that respiratory transmission of enteroviruses by droplets from the oral cavity may also explain the high secondary infection rate within households in some outbreaks. It has been suggested that the primary replication sites for enteroviruses could be in the oral cavity and/or gastrointestinal tract, with person-to-person transmission most commonly occurring via fecal–oral and oral–oral routes.[76] It has also been speculated that *Coxsackie A16* virus is spread through direct contact with the saliva, mucus, or feces of an infected person,[77] but to date there is no evidence supporting transmission by kissing.

In experimental animal models, *Coxsackie B virus* was recovered in the whole saliva of rabbits as early as 2 minutes after injection of virus into an ear vein and it was suggested that virus might be transmitted by saliva of viremic animals without infection of the oropharyngeal tissues.[78]

EV71 is an important public health problem in Asia as it spreads easily to close contacts[79] and may cause central nervous system involvement, serious illness, or death.[80] ECHOvirus 9 strains have been isolated from the saliva of a mother and daughter, both suffering acute salivary gland swelling, but with negative tests for mumps virus.[81]

The evidence suggests that *enteroviruses* could be transmitted through saliva or via kissing.

4.1.7 Hantaviruses

Hantaviruses are zoonoses, rodents being the natural hosts. *Hantaviruses* cause mainly hemorrhagic fever with renal syndrome (HFRS) and *Hantavirus* cardiopulmonary syndrome (HCPS) but there are other forms.

Transmission to humans is usually by inhalation of aerosolized virus-contaminated rodent excreta. *Hantavirus* RNA has also been detected in saliva of patients with HFRS[20] and from *Puumala hantavirus*–infected people.

Human-to-human transmission of *hantaviruses* appears rare, except in the case of *Andes virus* (ANDV).[82] It has been speculated that ANDV may be secreted into saliva, which might be a transmission route for ANDV between humans,[83] but there is no reliable evidence, and there is some evidence of salivary inhibitory factors.[84] *Hantaan hantavirus* appears to be inhibited by salivary mucin—though resistant to histatin 5, lysozyme, lactoferrin, and SLPI.[84]

A prospective study of household contacts of patients with HCPS concluded that transmission risks increased in people partaking in deep kissing or sex with an infected individual.[85] The evidence suggests that *hantaviruses* could be transmitted through deep kissing.

4.1.8 Hepatitis Viruses

There is a number of hepatotropic viruses, *Hepatitis A, B,* and *C viruses*, and others (e.g., EBV) which can cause hepatitis and may be present in saliva as well as blood and other body fluids, such as semen and gastric juices. Hepatitis A is typically transmitted predominantly by the oro-fecal route and causes a transient illness with hepatitis and no serious sequelae. Hepatitis B and C (formerly hepatitis non-A, non-B) virus infections however are transmitted mainly parenterally—blood transfusion and intravenous drug use being the most frequent risk factors, though sexual transmission may also occur. Illness may be prolonged with chronic carriage and can also cause complications such as cirrhosis and hepatocellular carcinoma.

4.1.8.1 Hepatitis A Virus (HAV)

HAV-RNA may be found in oral fluids[86] though it is present in low frequencies in saliva from HAV-infected patients[21] while others failed to detect it in saliva obtained during a *Hepatitis A virus* outbreak.[87] It seems that saliva can be a potential route of transmission.

4.1.8.2 Hepatitis B Virus (HBV)

HBV-DNA has been detected in saliva from viremic HBV-infected subjects.[88] Transmission has been demonstrated by subcutaneous inoculation of HBV-infected saliva into animals,[89] but the role of saliva in the person-to-person transmission of HBV infection is still not completely defined. Hepatitis B surface antigen (HBsAg) was found in saliva of 76% of patients with severe hepatitis and in 81% of chronic carriers.[90] It has been implied that transmission may be directly from mouth-to-mouth by kissing or by the exchange of saliva on chewed toys and candies[90] but no infections have been confirmed in susceptible persons with an intact oral mucosa who were orally exposed to HBV-infected saliva[91] or students who had oral exposures to HBsAg-positive saliva via contaminated musical instruments.[92] HBV has been transmitted by a human bite,[93] and a case of HBV transmission has been reported in a person who developed hepatitis B presumably after deep kissing with his partner.[94]

However, saliva has a HBV load 1000–10,000 times lower than blood, so the possible transmission of HBV by contaminated saliva remains low.[89,95-97] There are no reports of HBV transmission during mouth-to-mouth ventilation or cardiopulmonary resuscitation (CPR) training with a mannequim.[98-100]

The evidence therefore points to the possible transmission of HBV via saliva being low.

4.1.8.3 Hepatitis C Virus (HCV)

HCV is transmitted primarily through blood-to-blood contact though 62% of chronic carriers may have HCV-positive saliva.[101-103] HCV-RNA can be detected in the saliva of HCV-infected patients,[23] which might provide an argument for the possible transmission of HCV via saliva, but the salivary HCV viral load is significantly lower than the blood viral load.[104] HCV-RNA in saliva is associated with the level of serum viral load but not with periodontal or liver disease severity.[105] No case of transmission by HCV after mouth-to-mouth ventilation has been described.[106] Transmission of HCV in CPR using CPR mannequins has not been reported even when exposed to HCV-contaminated saliva.[107]

Saliva may therefore be a source of occasional transmission of HCV,[108] particularly where there is deep kissing and there are oral mucosal lesions[109] though epidemiological studies suggest that the infective capacity of HCV in saliva is only low.[110]

4.1.8.4 Other Hepatotropic Viruses

Other hepatotropic viruses, such as *Hepatitis G virus* (HGV), have been found in saliva both from HGV-infected patients[25] and from non-hepatitis patients with oral diseases,[24] but transmission via saliva has not been demonstrated.

4.1.9 Herpesviruses

Human herpesvirus (HHV) infections are common, seen especially in younger people subclinically or producing fever and mucocutaneous lesions, and then remain latent but can be reactivated if immunity wanes. Many can be oncogenic. Oral disease associations examined have ranged from carcinogenesis[111,112] to periodontitis.[113]

4.1.9.1 Herpes Simplex Virus Type 1 (HSV-1; HHV-1)

Congenital HSV infection may result in fetal abnormalities in the TORCH (Toxoplasmosis, Other agents, Rubella, *Cytomegalovirus*, and *Herpes simplex*) syndrome. HSV-1 is typically acquired early in life, usually from direct contact with infected saliva or skin vesicles.[114] HSV can cause oral lesions mainly as gingivostomatitis, with recurrences usually as herpes labialis. *HSV type 1* DNA can be found in most cases of recurrent herpes labialis, both before and after the appearance of clinical lesions.[27] Furthermore, viral shedding with viral loads sufficient to be transmitted is more frequent than previously thought, even in otherwise healthy HSV-seronegative individuals.[26] HSV ulceration and recurrences and almost certainly shedding are increased and often more extensive in immunocompromised people such as those with HIV/AIDS or post-transplantation and after oral surgical procedures.[115]

HSV can survive up to 88 hours in dry gauze and 1.5 hours on hard surfaces[115] and has the potential to be spread by fomites.[116] From 2% to 10% of adults without clinical signs of disease have HSV-1 in their saliva.[117,118] HSV-1 transmission via saliva is thought to be common and though evidence is sparse it has been transmitted by mouth-to-mouth ventilation.[119–121] However, it is more likely that uninfected adults may contract HSV by kissing[122] or sexual practices but there is little hard evidence to confirm this.

4.1.9.2 Herpes Simplex Virus Type 2 (HSV-2; HHV-2)

HSV-2 typically causes anogenital herpetic lesions and infection is usually transmitted sexually. As with HSV-1, HSV-2 is capable of causing both anogenital and oral disease. Data from a limited number of studies indicate that HSV-2 shedding in saliva is uncommon and symptomless in healthy individuals[28] and usually occurs in the setting of simultaneous anogenital involvement.[123] Men who have sex with men and HIV-positive persons have slightly higher rates of oral HSV-2 shedding than do otherwise healthy individuals.[124,125] However, some studies showed an incidence of HSV-2 in saliva of HIV-infected patients as low as that detected in controls.[126] On the contrary, high HSV-2 salivary detection rates have been reported among Brazilian HIV-infected and healthy children (4.2% and 8.3%, respectively).[127] HSV-2 may not be detected in the saliva of HIV-seropositive persons undergoing highly active antiretroviral therapy.[128] Based on a study on *herpesvirus* prevalence among 16-year-old Swedish girls, it was suggested that "transmission of *herpesviruses* is common in adolescence, and sex, even with regard to its close association with kissing, is one important determinant"[129]; however, in that series antibodies to HSV-2 were seen only in 1% of participants, and to date kissing transmission has not been reliably demonstrated.

4.1.9.3 Varicella Zoster Virus (VZV; HHV-3)

VZV causes mainly varicella (chickenpox) and zoster (shingles) but there can be more serious complications including meningoencephalitis, and occult forms of VZV-induced disease, including zoster sine herpete and enteric zoster. VZV lesions and recurrences are increased and often more extensive in immunocompromised people such as those with HIV/AIDS or posttransplantation.[130] Chickenpox is highly contagious—spread is mainly airborne.[131] VZV-DNA has also been found in the saliva of patients with clinical VZV infections, being detectable before just the varicella rash appears and for 1–2 weeks thereafter.[132] VZV can also be found in saliva in zoster[48] and VZV meningoencephalitis.[133] Saliva samples from individuals aged older than 60 years show VZV in those with a history of zoster, as well as in some healthy older controls with no history of zoster.[134] VZV may also be found in saliva after VZV immunization.[135] Nevertheless, it has been suggested that the detection of VZV-DNA in saliva may be useful in the diagnosis of atypical cases of varicella, zoster sine herpete, neurological syndromes when cerebrospinal fluid is not available and in Bell's palsy, and atypical pain syndromes.[132] It has been suggested that stress favors VZV shedding in saliva, as was shown in a group of asymptomatic astronauts, both during and after space flights.[47] VZV is not detected in saliva from healthy adults.[136] Though there is little reliable evidence of the potential for VZV transmission via saliva, it is highly likely in the scenarios discussed earlier.

4.1.9.4 Epstein–Barr Virus (EBV; HHV-4)

Primary EBV infection in young children is usually subclinical. EBV clinical infection—infectious mononucleosis (glandular fever)—is seen mainly in children older than 10 years, adolescents, and young adults,[137] who present mainly with lymphadenopathy and sore throat.

Patients with infectious mononucleosis can shed EBV in saliva for months.[138] EBV-DNA loads in saliva during convalescence are high and associated with continued infectivity.[139] There is a high prevalence of EBV in saliva and throat washings from healthy children and adults in some geographical regions[18] and by adulthood, at least 90% of all people are seropositive to EBV. EBV may particularly be found in saliva when it is reactivated in immunocompromised people such as those with HIV/AIDS or posttransplantation[140,141] and in oral hairy leukoplakia.[142]

EBV replicates in oropharyngeal epithelial cells and can be spread through saliva.[143] EBV can be transmitted by kissing; prospective epidemiological studies in undergraduate university students have confirmed salivary transmission of EBV[19] though some authors criticized the study methods used.[144] EBV is also called "the Kissing Virus."[145] One anecdote is of a man with infectious mononucleosis who, for 12 hours, had shared a train carriage compartment with a women whom he had not met before and whom he never saw again but during the time in the train, they had repeatedly kissed intimately and some weeks later she developed infectious mononucleosis.[146] A large EBV viral load may be acquired during sexual intercourse[147] and presumably it can also be transmitted from carriers who cough.[144]

Even so, despite the lack of direct evidence, EBV transmission is most likely from saliva, especially by kissing.[19]

4.1.9.5 Cytomegalovirus (CMV; HHV-5)

CMV is another cause of "glandular fever" seen mainly before adulthood, and most adults have been previously infected with CMV but the virus may be reactivated in patients with immunocompromising conditions, e.g., HIV infection[148] and posttransplantation.[149] Multiple CMV strains are recognized and, in men who have sex with men coinfected with HIV, infections with several strains may be seen.[150] Infection by CMV B groups,[151] particularly CMV glycoprotein gB1 subtype (gB1) appear to cause most morbidity.[151,152]

CMV is also one of the most important known viral causes of fetal abnormalities (TORCH syndrome). CMV shedding in body fluids is seen in TORCH and also increased where there is reactivation in immunocompromised people.[141] CMV-seropositive children, especially infants, may be a high infective risk to pregnant women, mainly via saliva. Transmission is common in day care centers.[153]

Real-time PCR of saliva is the gold-standard diagnostic test for detection of CMV.[14] As well as CMV shedding in body fluids in immunocompromised people, apparently healthy but CMV-seropositive children can also shed CMV at high levels in saliva for months, highlighting the potentially high transmission risks posed by saliva,[13] probably the principal postnatal transmission route.[154]

Although there is no robust evidence of a risk of CMV transmission associated with kissing, this is a highly probable route.[155]

4.1.9.6 Human Herpesvirus 6 (HHV-6)

HHV-6 can cause a range of diseases from exanthema subitum (roseola infantum), mononucleosis syndromes, and pneumonitis to encephalitis.[156] Following primary infection, viral genomes may persist in peripheral blood and saliva of most apparently healthy individuals.[30,157] Serological studies indicate that HHV-6 infects most children by age 2 years and that older siblings serve as a source of transmission.[29] HHV-6 has a number of forms—HHV-6, HHV-6B, and also CiHHV-6, which is chromosomally integrated[158,159] and can be reactivated.

HHV-6 may be present in saliva from healthy adults[31] but even immunosuppression by AIDS has little effect on HHV-6 shedding in saliva.[160]

HHV-6 transmission may occur via an oral route.[161]

4.1.9.7 Human Herpesvirus 7 (HHV-7)

HHV-7 infection is usually a benign and self-limited disease of childhood and rarely has complications though it may affect the central nervous system. It has been implicated in pityriasis rosea and, like HHV-6 it can also cause exanthem subitum.

HHV-7 DNA has been detected in saliva from practically all patients with pityriasis rosea[32] and frequently from healthy adults, particularly women, with regional variations.[31,162–164] HHV-7 is present in saliva in adults, and immunosuppression increases both frequency of detection and viral load.[160]

HHV-7 may well be transmitted person-to-person via saliva.[165]

4.1.9.8 Human Herpesvirus 8 (HHV-8; Kaposi Sarcoma Herpesvirus; KSHV)

HHV-8 (KSHV) can cause Kaposi's sarcoma (KS), primary effusion lymphoma, and multicentric Castleman's disease. KSHV infection is endemic in sub-Saharan Africa, the Mediterranean littoral, and China (Xinjiang region), but Western Europe and United States have a low prevalence. KSHV is found mainly in immunocompromised people such as those with HIV/AIDS or posttransplantation.

KSHV is detectable in more than 90% of KS lesions,[166,167] and oral replication is an essential feature of infection.[168]

In areas of high HIV prevalence, such as sub-Saharan Africa, the Mediterranean littoral, and the Xinjiang region in China, KSHV can be transmitted sexually or by contaminated blood transfusions and tissue transplants,[169] or via saliva contact.[170] KSHV is found in saliva from some 35%–80% of apparently healthy adults especially from certain geographical regions such as Brazil[33] and also particularly from immunocompromised people.[171] There is a high prevalence of KSHV in the saliva of patients with KS, even in the absence of intraoral lesions.[34]

There is consensus that saliva is the main route of KSHV transmission, especially in children in endemic areas,[172] that household exposure increases the risk for early childhood infection, and that specific feeding behaviors probably play a role in some cases.[173] Occupational transmission of KSHV to healthcare workers is uncommon.[174] However, epidemiological evidence points to particular sexual behaviors, including deep kissing, as being significant transmission risk factors for KSHV after childhood.[175]

4.1.10 Human Immunodeficiency Viruses (HIV)

Human Immunodeficiency Viruses cause HIV disease and the Acquired Immune Deficiency Syndrome (AIDS). The possibility of interpersonal HIV contagion by saliva is often still a common belief[176] and, e.g., some parents fear transmitting HIV to, or catching infection from their children through saliva.[177] However, some 3 decades ago, it was stated that there is no epidemiological evidence that HIV is transmitted by saliva,[178] and, although such transmission is theoretically possible[179] little has changed subsequently.

HIV can indeed be isolated from saliva, but infrequently.[180] Saliva may of course, also contain blood[181] and it has been suggested that HIV may be transmitted after deep kissing,[182] especially if there is blood in the mouth.[183] Hemoglobin concentrations in saliva are higher in HIV-positive intravenous drug abusers (IVDA) than in HIV-negative IVDA and controls, and in AIDS a mean of 1.3 μL blood/mL saliva has been estimated.[183] However, there is only a low level of HIV in saliva even when it is likely contaminated with blood.[184] Transmission to household nonsexual contacts is improbable.[185]

HIV transmission is unlikely with normal household contacts[186] and, although it has been stated by the Centers for Disease Control and Prevention (CDC) that deep kissing might result in viral transmission,[187] CDC now declares that HIV is not spread via saliva, considers open-mouth kissing a low risk for HIV transmission and yet recommends against engaging in this activity with known HIV-infected persons.[188] Nevertheless, CDC has apparently investigated only a single case attributed to contact with blood in the mouth.[189]

Mouth-to-mouth ventilation is not implicated in the transmission of HIV, even though contaminated saliva contacts open wounds.[190] Blood is visible in saliva or vomit in some resuscitations and this with oral microlesions in 50% of healthy rescuers, makes blood-to-blood contact possible.[109]

Nevertheless, even after exposure of healthcare workers to blood-contaminated saliva or blood through accidental needle stick injuries, the risk of infection has been consistently less than 1% and transmission in normal dental practice is improbable.[191]

4.1.11 Human Papillomaviruses (HPV)

HPV can cause benign warty lesions on oral and other epithelia, and some oncogenic types are implicated in cervical, anogenital, and oropharyngeal cancers. A number of viruses have been examined for possible roles in oral carcinogenesis.[112,192–194] including HPV.[195]

A small but noteworthy proportion of clinically healthy individuals have oral HPV infections,[36,196,197] including with HPV types known to cause cancer in the oral region.[198]

One HPV reservoir may be in the gingivae.[199] HPV are often latent, lesions may be small or symptomless and often undetected, and virus may be reactivated by immunoincompetence.[200–204] HPV lesions are often more extensive in immunocompromised people such as those with HIV/AIDS or posttransplantation.[205]

Some years ago, it was suggested that oral cancer might be a sexually transmitted viral infection.[206,207] Early focus on *herpes simplex* virus, turned to HPV. Oropharyngeal cancer in particular has proved to be significantly associated with oral HPV type 16 (an oncogenic HPV) and associated with a high lifetime number of vaginal-sex partners and oral-sex partners with oncogenic HPV-16 DNA being detected in many (72%) of cancers.[208] HPV-DNA in oral cancer biopsy specimens is more frequent among subjects who reported >1 sexual partner or who practiced oral sex.[209] Oropharyngeal cancer, tonsillar in particular, is increased in patients with anogenital cancer[210] and in females with cervical cancer and in their partners.[211] There have been significant increases in tonsil and base of tongue cancers in males, and base of tongue cancer in females with HPV-associated oropharyngeal cancer are increasing.[212]

HPV horizontal, nonsexual transmission may be responsible for oral infection in children.[213] Household transmission may occur via saliva and the shared use of contaminated objects.[37] Partners may carry the same oral HPV, and persistent oral HPV infection of the spouse increases the risk of persistent oral HPV infection 10-fold in the other spouse.[214] HPV is particularly common in those with oropharyngeal cancer,[215] though in one study partners of patients with HPV-OPC did not seem to have more oral HPV infection compared with the general population.[216]

Sexual behavior may impact on the risk of infection and though some authors consider kissing to be a low-risk activity,[214,217] deep (French) kissing has been associated with the development of oral HPV infection;[218] this finding has been confirmed in epidemiological studies on young adults of both genders[219] and men who have sex with men,[220] but not, in a study based on a questionnaire on sexual behavior of Australian university students, where no significant differences were found in the number of partners for deep kissing, between those with oral HPV infection and HPV-negative students.[221]

In summary, some studies imply therefore that oral HPV may be transmitted by deep kissing or by oral sex (mouth-to-genital or mouth-to-anus contacts), while others have not. The likelihood of contracting HPV from kissing or having oral sex with a person who carries HPV is not therefore, completely certain.[222]

4.1.12 Human Polyomaviruses

Humans may contract infections with *BK polyomavirus* (BKV), *JC polyomavirus* (JCV), or *Merkel cell polyomavirus* (MCV). *Polyomavirus* primary infections generally occur early in life and are implicated respectively in nephropathy, progressive multifocal leukoencephalopathy, and

Merkel cell carcinoma. The role of novel *Human polyomaviruses, KI* (KIV), and *WU* (WUV) is unclear.

Polyomavirus detection is generally highest among people 15–19 years of age; WUV infections being more frequent between those ages and decreasing later, but BKV excretion peaking and persisting during the third decade of life, and KIV is more common in subjects ≥ 50 years of age.[42] BKV, JCV, WUV, and KIV are found in the saliva of some healthy individuals[223] and may be transmitted by saliva.[42] *Polyomavirus* reactivation is more common in immunocompromised people and these viruses may be found in saliva from HIV-positive children[224] and in renal transplant recipients.[225]

MCV is widespread in the human body, and in saliva it is more common than in samples from the lung and genitourinary system, so transmission via saliva would seem possible,[43] and MCV may be acquired through close contact between young siblings and between mothers and their children.[226]

BKV has also been detected in saliva from apparently healthy individuals,[42] and salivary gland cells may be a site of virus replication,[227] supporting the hypothesis that saliva may be a route for BKV transmission. In contrast, JCV shedding in saliva is rare.[228]

Thus although saliva may be a route of transmission for some human *polyomaviruses*, the possibility that kissing is a significant risk activity remains to be clarified.

4.1.13 Influenza Viruses

There are three main types of *Human influenza viruses, types A* (IVA), *B*, and *C*. All cause severe lower respiratory disease. Nasopharyngeal colonization by *Streptococcus pneumoniae* (*pneumococcus*) is thought to be a prerequisite for developing influenza, and in some older people with influenza-like-illnesses *pneumococci* are plentiful in saliva.[229] Significant decreases in numbers of salivary anaerobic bacterial (CFUs), and neuraminidase and trypsin-like proteases levels after professional oral healthcare suggests that oral hygiene maintenance can be effective in reducing influenza.[230]

H1N1 (swine flu) and some other novel strains of *Influenza viruses* are carried by pigs, poultry, or other birds and several novel viruses have appeared in resource-rich countries mainly in traveling people.[52] *Avian influenza A (H5N1)* and *A (H7N9) viruses* which circulate widely in some poultry populations and sporadically infect humans include: in eastern China the *H7N9 influenza A virus*, and in the United States a *Swine-like influenza H3N2 variant virus*. Sporadic human cases of avian *A (H5N6), A (H10N8)*, and *A (H6N1)* have also emerged.[231]

Saliva in people with influenza may well contain the virus. About half of the patients with influenza have positive saliva and nasopharyngeal swabs within 24 hours from the onset of symptoms.[232] Saliva sampling for H1N1 is accurate, reliable, and more convenient than a nasopharyngeal swab.[38]

It seems that saliva represents an important initial barrier to *Influenza virus* infection. Consequently, it is unlikely that kissing could constitute a route for transmission.

4.1.14 Measles Virus
The *Measles virus* causes measles, a highly contagious disorder, usually in children manifesting with fever and rash sometimes with more serious complications.[233] Outbreaks of measles communities such as in Gypsy-Travelers are well recognized.[234]

Measles virus is found in saliva,[45,235] where nucleic acid can be amplified by PCR[39] and directly from using a point-of-care test.[236] Measles is usually spread by droplets but kissing may well also transmit the virus.

4.1.15 Metapneumovirus (hMPV)
Human metapneumovirus (hMPV) is a respiratory pathogen that can cause features ranging from asymptomatic infection to severe bronchitis, mainly but by no means exclusively in children, and is therefore clinically similar to infections with *Respiratory syncytial virus, Parainfluenza virus type 1*, and *Human parainfluenza virus type 3.*

hMPV is found in saliva, more in children within 3 days of onset of symptoms than later.[3,237]

Although we found no reliable evidence of the virus transmission by saliva, it is highly likely.

4.1.16 Molluscum Contagiosum Virus

Molluscum contagiosum is a benign *poxvirus* infection of the skin or very occasionally the mucosae. Lesions are seen mainly in children, sexually active adults, and those who are immunocompromised.

We found no reliable evidence of *Molluscum contagiosum virus* in, or transmission by, saliva.

4.1.17 Mumps Virus

Mumps virus typically causes acute sialadenitis (usually parotitis) but is a systemic infection with a variety of possible extra-salivary complications.[238]

Mumps transmission has occurred despite prompt isolation of cases after the onset of parotitis, indicating viral shedding before the onset of parotitis.[239] Mumps transmission can occur from persons with subclinical or clinical infections and during the prodromal or symptomatic phases of illness and within the subsequent 5 days.[240] The CDC, American Academy of Pediatrics (AAP), and Healthcare Infection Control Practices Advisory Committee (HICPAC) recommend a 5-day period after the onset of parotitis, both for isolation of persons with mumps in either community or healthcare settings and for use of standard infection control precautions and droplet precautions.[240]

Mumps virus is transmitted mainly by respiratory droplets but virus can also be found in saliva.[45] *Mumps virus* can be isolated from saliva and throat swabs from 7 days before to 8 days after the onset of parotitis, isolation rates are much greater closer to parotitis onset, the viral load decreasing substantially over the first 4 days after illness onset and becoming extremely low thereafter. The raised salivary load of mumps virus suggests a risk for transmission.[241]

Although we found no reliable evidence of mumps virus transmission by saliva, it is highly likely.

4.1.18 Nipah Virus

Nipah virus (NiV) is a *paramyxovirus*, whose main reservoir host is the fruit bat, first identified in Malaysia and Singapore during an outbreak of encephalitis and respiratory illness in farmers and people with close pig contacts. It is potentially lethal. Though uncommon, person-to-person transmission of NiV may occur mainly via respiratory secretions and body fluids, including saliva.[41,242–244]

Although we found no reliable evidence of NiV transmission by saliva, it is highly likely.

4.1.19 Noroviruses

Noroviruses (NoV or human Nov (HuNov)) are *calciviruses* which commonly cause acute gastroenteritis (Winter vomiting disease). NoV genogroup I (GI) (includes *Desert Shield virus, Norwalk virus,* and *Southampton virus*) or genogroup II (GII) (includes *Bristol virus, Hawaii virus, Lordsdale virus, Mexico virus, Snow Mountain virus* and *Toronto virus*) causes most infections.[245] Deaths are usually in the very young, old, or immunosuppressed and are rare in resource-rich communities.[246]

Differences in NoV susceptibility relate to factors including histo-blood group antigens (HBGAs) (i.e., the ABO blood group, the Lewis phenotype, and the secretor status), FUT2 (FUcosylTransferase 2), and FUT3 genotypes.[247,248]

Noroviruses are extremely contagious,[249] epidemics being seen mainly in closed communities such as cruise ships and long-stay facilities, and some research suggests as few as five virus particles are enough to transmit infection.[250] NoV can survive for long periods outside a host depending on the surface and temperature conditions. One study found NoV on surfaces used for food preparation 7 days after contamination.[251] It can also survive for months in contaminated water, weeks on hard surfaces, and up to 12 days on fabrics.[252]

There can be feco-oral transmission of NoV but salivary transmission appears poorly documented. It has been suggested that it is impossible to get infected by kissing someone who is not yet showing symptoms. However, it may be possible to catch it from someone who has recently vomited by kissing them, as viral particles may be in their mouth from vomitus.[253]

4.1.20 Parainfluenza Viruses (HPIVs)

Human parainfluenza viruses (HPIVs) are RNA viruses in a group of four distinct serotypes, which commonly cause respiratory illnesses in infants and young children but anyone can suffer HPIV illness of fever, runny nose, and cough. Furthermore, HPIVs can also cause more severe illness, such as croup or pneumonia.

Most children have been infected by HPIV-3 by age 2 years and by HPIV-1 and 2 by age 5, and HPIVs are found in saliva.[3]

Transmission is mainly respiratory but probably also by saliva.

4.1.21 Parvoviruses

Parvovirus B19 infection may cause erythema infectiosum—a mild fever and rash with oral erythema, and sometimes arthralgia or arthritis. However, severe outcomes of *Parvovirus B19* infection may occur. During pregnancy, *Parvovirus B19* infection of the fetus can cause fetal loss in the first trimester, or extensive hemolysis. In patients with hemolytic anemias, *Parvovirus B19* can produce a transient aplastic crisis. In infants, or in immunocompromised patients, *Parvovirus B19* can cause serious hemolysis.

Parvovirus transmission is mainly via respiratory droplets[254] but could be through saliva.

4.1.22 Rabies Virus

Rabies virus expands worldwide, especially in Asia and Africa. *Rabies virus* can be detected in saliva of rabies patients[255] now with RT-PCR.[44]

Rabies virus is present in fluids and tissues during the first 5 weeks of transmission, but there are few well-documented reports of human-to-human transmission—and these invariably are in corneal[256] or organ transplant recipients.[257]

Although the infection has never been well-documented, human-to-human transmission of rabies following saliva exposure remains at least a theoretical possibility.

4.1.23 Respiratory Syncytial Virus (RSV)

RSV is a common cause of respiratory illness indistinguishable from the common cold, and seen mainly in young children. It may lead to lower respiratory disease. Secretion of blood group antigens is associated with respiratory virus diseases.[258]

RSV can be found in saliva[3,237] and can remain viable for at least 30 minutes on hands or for up to 5 hours on surfaces.

Although we found no reliable evidence of RSV transmission by saliva, it is highly likely.

4.1.24 Rhinoviruses

Rhinoviruses are a frequent cause of the common cold.[259] Transmission is mainly via droplets, but viruses may persist in moist secretions on fomites.

Communicability of *rhinoviruses* showed transmission between partners of 41% and 33% for types 16 and 55, respectively,[260,261] but transmission was rare unless the donor spent hours with the partner and had virus on their hands and anterior nares.

Saliva in adults contains neutralizing antibodies to *rhinoviruses*.[262] Few *rhinovirus* particles survive in saliva and therefore, though transmission via saliva must be possible, normal kissing is not the usual mode of infection spread.

4.1.25 Rotaviruses

Rotaviruses are (with *noroviruses*) the most important causes of acute gastroenteritis, mainly in children. *Rotaviruses* are highly contagious.

The host secretor status (FUT2 genotype) affects the expression of HBGAs which act as sites for viral attachment to the gastrointestinal, respiratory, and genitourinary tract epithelia.[263] *Rotavirus* VP4 spike protein (VP8*) engages sialic acid in the viral binding to cellular receptors, facilitating viral attachment and entry.[264] In newborns and infants immunized against *rotavirus*, serum and saliva IgA antibodies conceivably assist in protection.[265]

Rotaviruses can be found in saliva and other body fluids in infected patients,[266] presumably transmitting infection.

4.1.26 Rubella Virus

Rubella virus causes infection mainly in children who may be symptomless or develop a rash, fever, and occasionally other symptoms or signs including lymphadenopathy and palatal purpura. Congenital rubella infection may result in TORCH syndrome.

In rubella, the virus RNA can be found in saliva.[45,267] Although we found no reliable evidence of rubella transmission by saliva, it is highly likely.

4.1.27 Torque Teno Viruses (TTV)

TTV was first identified in a patient with non-A-E hepatitis but the viruses are now known to be ubiquitous,[268] with >90% of adults world-wide infected. It has been suggested that TTV infection is associated with many diseases, including some oral disease,[24] with little evidence.

TTV may be detected in saliva.[46] TTV transmission by saliva, though highly likely appears unsupported by reliable evidence.

4.1.28 Yellow Fever Virus

Yellow Fever virus is, like Dengue and Zika, a *flavivirus* transmitted by mosquitoes, usually *A. aegypti*, and is endemic to tropical regions of Africa and the Americas, causing potentially lethal hemorrhagic fever.[269] In Africa, yellow fever occurs in 34 countries and an epidemic in Angola in 2016 caused serious concern and was spread by travelers to at least China, Democratic Republic of Congo, Kenya, and Morocco. At least half of severely affected patients who do not receive treatment die within 14 days.

No human–human transmission of *Yellow* fever virus by saliva has been reported, and we found no reliable evidence of salivary transmission.

4.1.29 West Nile Virus (WNV)

WNV, a *flavivirus*, naturally maintained in a cycle between birds and mosquitoes, with occasional spillover by mosquito bites to humans, since isolated first in Uganda, has spread widely including the United States and Europe.[270] As with so many viruses, most WNV infections are asymptomatic, but there is a risk of potentially lethal neurological disease.

Although a case of possible sexual transmission of WNV has been reported,[271] we found no reliable evidence of WNV transmission by saliva.

4.1.30 Zika Virus (ZIKV)

ZIKV infection is a mosquito-borne, *flavivirus* disease associated mainly with Guillain-Barre syndrome and fetal microcephaly.[272] The magnitude of the current ZIKV epidemic has led to a declaration of a Public Health Emergency of International Concern by the WHO.[273]

ZIKV RNA has been detected in saliva from patients with Zika fever.[49] It has been suggested that the rate of ZIKV detection in saliva samples is higher even than in blood or urine.[49] Saliva and urine samples present higher viral load than serum.[49,274,275] Viral RNA is prolonged shedding in saliva for some weeks after symptom onset—with a slightly longer persistence time than in urine,[274,276] as it has previously been demonstrated for other vector-borne *flaviviruses*. In consequence, saliva has been recommended to test for ZIKV when drawing blood or processing of blood samples may be difficult.[275,276]

Nonvector-borne ZIKV transmission plays a role in the spread of ZIKV.[277] One American scientist contracted Zika while working in Senegal in 2008, and transmitted ZIKV infection to his wife; a sexual transmission of the virus was suggested, but other possibilities such as exchange of other bodily fluids including saliva could not be ruled out; moreover, aphthous-type oral ulcers were observed in both spouses.[278] Here as yet no reliable evidence to support ZIKV transmission through human saliva,[9,279,280] but cytopathic effects have been reported when bringing saliva in contact with Vero cells, which suggests infectious potential.[277]

In a case report of ZIKV infection in a 24-year-old woman who was living in Paris and reported sexual contact with a man who had recently returned from Brazil, the authors could not rule out the possibility that transmission occurred through saliva exchanged through deep kissing,[281] but up to date we found here no reliable evidence of Zika transmission by kissing.

4.2 CLOSING REMARKS AND PERSPECTIVES

The transfer of body fluids might transmit viral infections, many of which are seen especially in immunocompromised people and some can be life-changing or even lethal, as it has happened in a number of catastrophes where viral agents such as *Human Immunodeficiency Viruses* and *hepatitis viruses* were transmitted by blood. Any risks of transmission from saliva appear not well-defined, and reliable evidence for virus transmission through kissing is sparse. Nevertheless, this route may be one where the *flaviviruses* at least may well spread. In this regard, the number of emerging viral diseases has increased dramatically in recent decades. In an ever more global society, the arrival of immigrants requires us to maximise universal barrier measures, in particular to avoid the transmission of pathogenic viruses not recognised by our immune system. Detection of some of these viruses in saliva, such as *Ebola virus* or *Zika virus* represent a new challenge for prevention of human to human transmission. Efforts are being made regarding the development of strategies for virus detection _including polymerase chain reaction-based methods, paper-based synthetic gene networks, immunoassays, magnetic nanoparticles-based assays or liposome-based detection assays. However, there is still a need to search for and improve upon more sensitive and specific detection methods for these challenging viruses. Substantial numbers oral viruses are shared amongst genetically unrelated, cohabitating individuals; most of these viruses are bacteriophages and their distribution over time within households indicates that they are frequently transmitted between the microbiomes of household contacts.

REFERENCES

1. Gautret P. Religious mass gatherings: connecting people and infectious agents. *Clin Microbiol Infect* 2015;**21**(2):107−8.

2. Corstjens PL, Abrams WR, Malamud D. Saliva and viral infections. *Periodontology 2000* 2016;**70**(1):93−110.

3. Robinson JL, Lee BE, Kothapalli S, Craig WR, Fox JD. Use of throat swab or saliva specimens for detection of respiratory viruses in children. *Clin Infect Dis* 2008;**46**(7):e61−4.

4. Edwards S, Carne C. Oral sex and the transmission of viral STIs. *Sex Transm Infect* 1998;**74**(1):6−10.

5. Fredricks DN, Relman DA. Sequence-based identification of microbial pathogens: a reconsideration of Koch's postulates. *Clin Microbiol Rev* 1996;**9**(1):18−33.

6. Porter SR, Mutlu S, Scully CM. Viral infections affecting periodontal health. *Periodontal Clin Investig* 1993;**15**(1):17−24.

7. Mikola H, Waris M, Tenovuo J. Inhibition of herpes simplex virus type 1, respiratory syncytial virus and echovirus type 11 by peroxidase-generated hypothiocyanite. *Antiviral Res* 1995;**26**(2):161−71.

8. Malamud D, Abrams WR, Barber CA, Weissman D, Rehtanz M, Golub E. Antiviral activities in human saliva. *Adv Dent Res* 2011;**23**(1):34−7.

9. Scully C, Robinson A. Check before you travel: Zika virus—another emerging global health threat. *Br Dent J* 2016;**220**(5):265−7.

10. Gardner J, Rudd PA, Prow NA, et al. Infectious chikungunya virus in the saliva of mice, monkeys and humans. *PLoS One* 2015;**10**(10):e0139481.

11. Goh G, Dunker A, Uversky V. Prediction of intrinsic disorder in MERS-CoV/HCoV-EMC supports a high oral-fecal transmission. *PLoS Curr* 2013;5.

12. Wang WK, Chen SY, Liu IJ, et al. Detection of SARS-associated coronavirus in throat wash and saliva in early diagnosis. *Emerg Infect Dis* 2004;**10**(7):1213−19.

13. Cannon MJ, Stowell JD, Clark R, et al. Repeated measures study of weekly and daily cytomegalovirus shedding patterns in saliva and urine of healthy cytomegalovirus-seropositive children. *BMC Infect Dis* 2014;**14**:569.

14. Pinninti SG, Ross SA, Shimamura M, et al. Comparison of saliva PCR assay versus rapid culture for detection of congenital cytomegalovirus infection. *Pediatr Infect Dis J* 2015;**34**(5): 536–7.

15. Andries AC, Duong V, Ly S, et al. Value of routine dengue diagnostic tests in urine and saliva specimens. *PLoS Negl Trop Dis* 2015;**9**(9). e0004100.

16. Bausch DG, Towner JS, Dowell SF, et al. Assessment of the risk of Ebola virus transmission from bodily fluids and fomites. *J Infect Dis* 2007;**196**(Suppl. 2):S142–7.

17. Graves PM, Rotbart HA, Nix WA, et al. Prospective study of enteroviral infections and development of beta-cell autoimmunity. Diabetes autoimmunity study in the young (DAISY). *Diabetes Res Clin Pract* 2003;**59**(1):51–61.

18. Ikuta K, Satoh Y, Hoshikawa Y, Sairenji T. Detection of Epstein–Barr virus in salivas and throat washings in healthy adults and children. *Microb Infect* 2000;**2**(2):115–20.

19. Balfour Jr HH, Odumade OA, Schmeling DO, et al. Behavioral, virologic, and immunologic factors associated with acquisition and severity of primary Epstein–Barr virus infection in university students. *J Infect Dis* 2013;**207**(1):80–8.

20. Pettersson L, Klingstrom J, Hardestam J, Lundkvist A, Ahlm C, Evander M. Hantavirus RNA in saliva from patients with hemorrhagic fever with renal syndrome. *Emerg Infect Dis* 2008;**14**(3):406–11.

21. Joshi MS, Bhalla S, Kalrao VR, Dhongade RK, Chitambar SD. Exploring the concurrent presence of hepatitis A virus genome in serum, stool, saliva, and urine samples of hepatitis A patients. *Diagn Microbiol Infect Dis* 2014;**78**(4):379–82.

22. Arora G, Sheikh S, Pallagatti S, Singh B, Singh VA, Singh R. Saliva as a tool in the detection of hepatitis B surface antigen in patients. *Compend Contin Educ Dent* 2012;**33**(3):174–6.

23. Hermida M, Ferreiro MC, Barral S, Laredo R, Castro A, Diz Dios P. Detection of HCV RNA in saliva of patients with hepatitis C virus infection by using a highly sensitive test. *J Virol Methods* 2002;**101**(1-2):29–35.

24. Yan J, Chen LL, Lou YL, Zhong XZ. Investigation of HGV and TTV infection in sera and saliva from non-hepatitis patients with oral diseases. *World J Gastroenterol* 2002;**8**(5):857–62.

25. Seemayer CA, Viazov S, Philipp T, Roggendorf M. Detection of GBV-C/HGV RNA in saliva and serum, but not in urine of infected patients. *Infection* 1998;**26**(1):39–41.

26. Miller CS, Danaher RJ. Asymptomatic shedding of herpes simplex virus (HSV) in the oral cavity. *Oral Surg Oral Med Oral Pathol Oral Radiol Endod* 2008;**105**(1):43–50.

27. Gilbert SC. Oral shedding of herpes simplex virus type 1 in immunocompetent persons. *J Oral Pathol Med* 2006;**35**(9):548–53.

28. Tateishi K, Toh Y, Minagawa H, Tashiro H. Detection of herpes simplex virus (HSV) in the saliva from 1,000 oral surgery outpatients by the polymerase chain reaction (PCR) and virus isolation. *J Oral Pathol Med* 1994;**23**(2):80–4.

29. Zerr DM, Meier AS, Selke SS, et al. A population-based study of primary human herpesvirus 6 infection. *N Engl J Med* 2005;**352**(8):768–76.

30. Leibovitch EC, Brunetto GS, Caruso B, et al. Coinfection of human herpesviruses 6A (HHV-6A) and HHV-6B as demonstrated by novel digital droplet PCR assay. *PLoS One* 2014;**9**(3):e92328.

31. Magalhaes IM, Martins RV, Cossatis JJ, et al. Detection of human herpesvirus 6 and 7 DNA in saliva from healthy adults from Rio de Janeiro, Brazil. *Mem Inst Oswaldo Cruz* 2010;**105**(7):925–7.

32. Watanabe T, Kawamura T, Jacob SE, et al. Pityriasis rosea is associated with systemic active infection with both human herpesvirus-7 and human herpesvirus-6. *J Invest Dermatol* 2002;**119**(4):793–7.

33. de Souza VA, Sumita LM, Nascimento MC, et al. Human herpesvirus-8 infection and oral shedding in Amerindian and non-Amerindian populations in the Brazilian amazon region. *J Infect Dis* 2007;**196**(6):844−52.

34. Vieira J, Huang ML, Koelle DM, Corey L. Transmissible Kaposi's sarcoma-associated herpesvirus (human herpesvirus 8) in saliva of men with a history of Kaposi's sarcoma. *J Virol* 1997;**71**(9):7083−7.

35. Navazesh M, Mulligan R, Kono N, et al. Oral and systemic health correlates of HIV-1 shedding in saliva. *J Dent Res* 2010;**89**(10):1074−9.

36. Kreimer AR, Bhatia RK, Messeguer AL, Gonzalez P, Herrero R, Giuliano AR. Oral human papillomavirus in healthy individuals: a systematic review of the literature. *Sex Transm Dis* 2010;**37**(6):386−91.

37. Lopez-Villanueva ME, Conde-Ferraez L, Ayora-Talavera G, Ceron-Espinosa JD, Gonzalez-Losa Mdel R. Human papillomavirus 13 in a Mexican Mayan community with multifocal epithelial hyperplasia: could saliva be involved in household transmission? *Eur J Dermatol* 2011;**21**(3):396−400.

38. Bilder L, Machtei EE, Shenhar Y, Kra-Oz Z, Basis F. Salivary detection of H1N1 virus: a clinical feasibility investigation. *J Dent Res* 2011;**90**(9):1136−9.

39. Oliveira SA, Siqueira MM, Camacho LA, Castro-Silva R, Bruno BF, Cohen BJ. Use of RT-PCR on oral fluid samples to assist the identification of measles cases during an outbreak. *Epidemiol Infect* 2003;**130**(1):101−6.

40. Royuela E, Castellanos A, Sanchez-Herrero C, Sanz JC, De Ory F, Echevarria JE. Mumps virus diagnosis and genotyping using a novel single RT-PCR. *J Clin Virol* 2011;**52**(4):359−62.

41. Luby SP, Gurley ES, Hossain MJ. Transmission of human infection with Nipah virus. *Clin Infect Dis* 2009;**49**(11):1743−8.

42. Robaina TF, Mendes GS, Benati FJ, et al. Shedding of polyomavirus in the saliva of immunocompetent individuals. *J Med Virol* 2013;**85**(1):144−8.

43. Loyo M, Guerrero-Preston R, Brait M, et al. Quantitative detection of Merkel cell virus in human tissues and possible mode of transmission. *Int J Cancer* 2010;**126**(12):2991−6.

44. Crepin P, Audry L, Rotivel Y, Gacoin A, Caroff C, Bourhy H. Intravitam diagnosis of human rabies by PCR using saliva and cerebrospinal fluid. *J Clin Microbiol* 1998;**36**(4):1117−21.

45. Jin L, Vyse A, Brown DW. The role of RT-PCR assay of oral fluid for diagnosis and surveillance of measles, mumps and rubella. *Bull World Health Organ* 2002;**80**(1):76−7.

46. Naganuma M, Tominaga N, Miyamura T, Soda A, Moriuchi M, Moriuchi H. TT virus prevalence, viral loads and genotypic variability in saliva from healthy Japanese children. *Acta Paediatr* 2008;**97**(12):1686−90.

47. Mehta SK, Cohrs RJ, Forghani B, Zerbe G, Gilden DH, Pierson DL. Stress-induced subclinical reactivation of varicella zoster virus in astronauts. *J Med Virol* 2004;**72**(1):174−9.

48. Mehta SK, Tyring SK, Gilden DH, et al. Varicella-zoster virus in the saliva of patients with herpes zoster. *J Infect Dis* 2008;**197**(5):654−7.

49. Musso D, Roche C, Nhan TX, Robin E, Teissier A, Cao-Lormeau VM. Detection of Zika virus in saliva. *J Clin Virol* 2015;**68**:53−5.

50. Liddle OL, Samuel MI, Sudhanva M, Ellis J, Taylor C. Adenovirus urethritis and concurrent conjunctivitis: a case series and review of the literature. *Sex Transm Infect* 2015;**91**(2):87−90.

51. Cao B, Huang GH, Pu ZH, et al. Emergence of community-acquired adenovirus type 55 as a cause of community-onset pneumonia. *Chest* 2014;**145**(1):79−86.

52. Gautret P, Gray GC, Charrel RN, et al. Emerging viral respiratory tract infections—environmental risk factors and transmission. *Lancet Infect Dis* 2014;**14**(11):1113−22.

53. Kawahira H, Matsushita K, Shiratori T, et al. Viral shedding after p53 adenoviral gene therapy in 10 cases of esophageal cancer. *Cancer Sci* 2010;**101**(1):289−91.

54. Katti R, Shahapur PR, Udapudi KL. Impact of chikungunya virus infection on oral health status: an observational study. *Indian J Dent Res* 2011;**22**(4):613.

55. Gerardin P, Barau G, Michault A, et al. Multidisciplinary prospective study of mother-to-child chikungunya virus infections on the Island of La Reunion. *PLoS Med* 2008;**5**(3):e60.

56. Rolph MS, Zaid A, Mahalingam S. Salivary transmission of the chikungunya arbovirus. *Trends Microbiol* 2016;**24**(2):86−7.

57. Zumla A, Hui DS, Al-Tawfiq JA, Gautret P, McCloskey B, Memish ZA. Emerging respiratory tract infections. *Lancet Infect Dis* 2014;**14**(10):910−11.

58. Gralinski LE, Baric RS. Molecular pathology of emerging Coronavirus infections. *J Pathol* 2015;**235**(2):185−95.

59. Menachery VD, Yount Jr BL, Sims AC, et al. SARS-like WIV1-CoV poised for human emergence. *Proc Natl Acad Sci USA* 2016;**113**(11):3048−53.

60. Collins AR, Grubb A. Cystatin D, a natural salivary cysteine protease inhibitor, inhibits coronavirus replication at its physiologic concentration. *Oral Microbiol Immunol* 1998;**13**(1): 59−61.

61. Mizuno Y, Kotaki A, Harada F, Tajima S, Kurane I, Takasaki T. Confirmation of dengue virus infection by detection of dengue virus type 1 genome in urine and saliva but not in plasma. *Trans R Soc Trop Med Hyg* 2007;**101**(7):738−9.

62. Poloni TR, Oliveira AS, Alfonso HL, et al. Detection of dengue virus in saliva and urine by real time RT-PCR. *Virol J* 2010;**7**:22.

63. Korhonen EM, Huhtamo E, Virtala AM, Kantele A, Vapalahti O. Approach to non-invasive sampling in dengue diagnostics: exploring virus and NS1 antigen detection in saliva and urine of travelers with dengue. *J Clin Virol* 2014;**61**(3):353−8.

64. Formenty P, Leroy EM, Epelboin A, et al. Detection of Ebola virus in oral fluid specimens during outbreaks of Ebola virus hemorrhagic fever in the Republic of Congo. *Clin Infect Dis* 2006;**42**(11):1521−6.

65. Towner JS, Rollin PE, Bausch DG, et al. Rapid diagnosis of ebola hemorrhagic fever by reverse transcription-PCR in an outbreak setting and assessment of patient viral load as a predictor of outcome. *J Virol* 2004;**78**(8):4330−41.

66. Scully C, Samaranayake L, Petti S, Nair RG. Infection control: ebola aware; ebola beware; ebola healthcare. *Br Dent J* 2014;**217**(12):661.

67. Vetter P, Fischer 2nd WA, Schibler M, Jacobs M, Bausch DG, Kaiser L. Ebola virus shedding and transmission: review of current evidence. *J Infect Dis* 2016;**214**(Suppl. 3):S177−84.

68. Wozniak-Kosek A, Kosek J, Mierzejewski J, Rapiejko P. Progress in the diagnosis and control of ebola disease. *Adv Exp Med Biol* 2015;**857**:19−24.

69. Lopez-Sanchez A, Guijarro Guijarro B, Hernandez Vallejo G. Human repercussions of foot and mouth disease and other similar viral diseases. *Med Oral* 2003;**8**(1):26−32.

70. Scott LA, Stone MS. Viral exanthems. *Dermatol Online J* 2003;**9**(3):4.

71. Chen SP, Huang YC, Li WC, et al. Comparison of clinical features between coxsackievirus A2 and enterovirus 71 during the enterovirus outbreak in Taiwan, 2008: a children's hospital experience. *J Microbiol Immunol Infect* 2010;**43**(2):99−104.

72. Osterback R, Vuorinen T, Linna M, Susi P, Hyypia T, Waris M. Coxsackievirus A6 and hand, foot, and mouth disease, Finland. *Emerg Infect Dis* 2009;**15**(9):1485−8.

73. Li W, Zhang X, Chen X, et al. Epidemiology of childhood enterovirus infections in Hangzhou, China. *Virol J* 2015;**12**:58.

74. Harvala H, Wolthers KC, Simmonds P. Parechoviruses in children: understanding a new infection. *Curr Opin Infect Dis* 2010;**23**(3):224−30.

75. Meenan PN, Hillary IB. Poliovirus in the upper respiratory tract of household contacts. *Lancet* 1963;**2**(7314):907−8.

76. Pallansch M, Roos R. Enteroviruses: polioviruses, coxsackieviruses, echoviruses, and newer enteroviruses. In: Knippe DM, Howley PM, editors. *Fields Virology.* 4th ed. Philadelphia, PA: Lippincott Williams and Wilkins; 2001. p. 723−75.

77. Rao PK, Veena K, Jagadishchandra H, Bhat SS, Shetty SR. Hand, foot and mouth disease: changing Indian scenario. *Int J Clin Pediatr Dent* 2012;**5**(3):220−2.

78. Madonia JV, Bahn AN, Calandra JC. Salivary excretion of coxsackie b-1 virus in rabbits. *Appl Microbiol* 1966;**14**(3):394−6.

79. Chang LY, Tsao KC, Hsia SH, et al. Transmission and clinical features of enterovirus 71 infections in household contacts in Taiwan. *JAMA* 2004;**291**(2):222−7.

80. Lee TC, Guo HR, Su HJ, Yang YC, Chang HL, Chen KT. Diseases caused by enterovirus 71 infection. *Pediatr Infect Dis J* 2009;**28**(10):904−10.

81. Wilterdink JB, Versteeg J, Kuiper A. Isolation of ECHO-9-viruses from the saliva of inflamed salivary glands. *Ned Tijdschr Geneeskd* 1959;**103**:1847−9.

82. Montgomery JM, Ksiazek TG, Khan AS. Hantavirus pulmonary syndrome: the sound of a mouse roaring. *J Infect Dis* 2007;**195**(11):1553−5.

83. Hardestam J, Lundkvist A, Klingstrom J. Sensitivity of Andes hantavirus to antiviral effect of human saliva. *Emerg Infect Dis* 2009;**15**(7):1140−2.

84. Hardestam J, Petterson L, Ahlm C, Evander M, Lundkvist A, Klingstrom J. Antiviral effect of human saliva against hantavirus. *J Med Virol* 2008;**80**(12):2122−6.

85. Ferres M, Vial P, Marco C, et al. Prospective evaluation of household contacts of persons with hantavirus cardiopulmonary syndrome in chile. *J Infect Dis* 2007;**195**(11):1563−71.

86. Mackiewicz V, Dussaix E, Le Petitcorps MF, Roque-Afonso AM. Detection of hepatitis A virus RNA in saliva. *J Clin Microbiol* 2004;**42**(9):4329−31.

87. Poovorawan Y, Theamboonlers A, Chongsrisawat V, Jantaradsamee P, Chutsirimongkol S, Tangkijvanich P. Clinical features and molecular characterization of hepatitis A virus outbreak in a child care center in thailand. *J Clin Virol* 2005;**32**(1):24−8.

88. Karayiannis P, Novick DM, Lok AS, Fowler MJ, Monjardino J, Thomas HC. Hepatitis B virus DNA in saliva, urine, and seminal fluid of carriers of hepatitis B e antigen. *Br Med J (Clin Res Ed)* 1985;**290**(6485):1853−5.

89. Bancroft WH, Snitbhan R, Scott RM, et al. Transmission of hepatitis B virus to gibbons by exposure to human saliva containing hepatitis B surface antigen. *J Infect Dis* 1977;**135**(1):79−85.

90. Villarejos VM, Visona KA, Gutierrez A, Rodriguez A. Role of saliva, urine and feces in the transmission of type B hepatitis. *N Engl J Med* 1974;**291**(26):1375−8.

91. Alter MJ. Epidemiology of hepatitis B in Europe and worldwide. *J Hepatol* 2003;**39**(Suppl. 1): S64−9.

92. Osterholm MT, Max BJ, Hanson M, Polesky HF. Potential risk of salivary-mediated viral hepatitis type B transmission from oral exposure to fomites. *J Hyg (Lond)* 1979;**83**(3):487−90.

93. Hui AY, Hung LC, Tse PC, Leung WK, Chan PK, Chan HL. Transmission of hepatitis B by human bite—confirmation by detection of virus in saliva and full genome sequencing. *J Clin Virol* 2005;**33**(3):254−6.

94. Kubo N, Furusyo N, Sawayama Y, et al. A patient in whom only hepatitis B virus (HBV) was thought to have been contracted, by kissing, from a same-sex partner coinfected with HBV and human immunodeficiency virus-1. *J Infect Chemother* 2003;**9**(3):260−4.

95. Krugman S, Giles JP, Hammond J. Infectious hepatitis. evidence for two distinctive clinical, epidemiological, and immunological types of infection. *JAMA* 1967;**200**(5):365–73.

96. Piazza M, Cacciatore L, Molinari V, Picciotto L. Letter: hepatitis B not transmissible via faecal-oral route. *Lancet* 1975;**2**(7937):706.

97. Scott RM, Snitbhan R, Bancroft WH, Alter HJ, Tingpalapong M. Experimental transmission of hepatitis B virus by semen and saliva. *J Infect Dis* 1980;**142**(1):67–71.

98. Centers for Disease Control and Prevention (CDC). Lack of transmission of hepatitis B to humans after oral exposure to hepatitis B surface antigen-positive saliva. *MMWR Morb Mortal Wkly Rep* 1978;**27**:247–8.

99. Glaser JB, Nadler JP. Hepatitis B virus in a cardiopulmonary resuscitation training course. risk of transmission from a surface antigen-positive participant. *Arch Intern Med* 1985;**145**(9):1653–5.

100. Osterholm MT, Bravo ER, Crosson JT, Polisky HF, Hanson M. Lack of transmission of viral hepatitis type B after oral exposure to HBsAg-positive saliva. *Br Med J* 1979;**2**(6200):1263–4.

101. Couzigou P, Richard L, Dumas F, Schouler L, Fleury H. Detection of HCV-RNA in saliva of patients with chronic hepatitis C. *Gut* 1993;**34**(2 Suppl.):S59–60.

102. Numata N, Ohori H, Hayakawa Y, Saitoh Y, Tsunoda A, Kanno A. Demonstration of hepatitis C virus genome in saliva and urine of patients with type C hepatitis: usefulness of the single round polymerase chain reaction method for detection of the HCV genome. *J Med Virol* 1993;**41**(2):120–8.

103. Puchhammer-Stockl E, Mor W, Kundi M, Heinz FX, Hofmann H, Kunz C. Prevalence of hepatitis-C virus RNA in serum and throat washings of children with chronic hepatitis. *J Med Virol* 1994;**43**(2):143–7.

104. Menezes GB, Pereira FA, Duarte CA, et al. Hepatitis C virus quantification in serum and saliva of HCV-infected patients. *Mem Inst Oswaldo Cruz* 2012;**107**(5):680–3.

105. Sosa-Jurado F, Hernandez-Galindo VL, Melendez-Mena D, et al. Detection of hepatitis C virus RNA in saliva of patients with active infection not associated with periodontal or liver disease severity. *BMC Infect Dis* 2014;**14**:72.

106. Arend CF. Transmission of infectious diseases through mouth-to-mouth ventilation: evidence-based or emotion-based medicine?. *Arq Bras Cardiol* 2000;**74**(1):86–97.

107. Perras ST, Poupard JA, Byrne EB, Nast PR. Lack of transmission of hepatitis non-A, non-B by CPR manikins. *N Engl J Med* 1980;**302**(2):118–19.

108. Xavier Santos RL, de Deus DM, de Almeida Lopes EP, Duarte Coelho MR, de Castro JF. Evaluation of viral load in saliva from patients with chronic hepatitis C infection. *J Infect Public Health* 2015;**8**(5):474–80.

109. Piazza M, Chirianni A, Picciotto L, Guadagnino V, Orlando R, Cataldo PT. Passionate kissing and microlesions of the oral mucosa: possible role in AIDS transmission. *JAMA* 1989;**261**(2):244–5.

110. Ferreiro MC, Dios PD, Scully C. Transmission of hepatitis C virus by saliva? *Oral Dis* 2005;**11**(4):230–5.

111. Eglin RP, Scully C, Lehner T, Ward-Booth P, McGregor IA. Detection of RNA complementary to herpes simplex virus in human oral squamous cell carcinoma. *Lancet* 1983;**2**(8353):766–8.

112. Scully C. Viruses and cancer: herpesviruses and tumors in the head and neck. A review. *Oral Surg Oral Med Oral Pathol* 1983;**56**(3):285–92.

113. Zhu C, Li F, Wong MC, Feng XP, Lu HX, Xu W. Association between herpesviruses and chronic periodontitis: a meta-analysis based on case-control studies. *PLoS One* 2015;**10**(12). e0144319.

114. Stuart-Harris C. The epidemiology and clinical presentation of herpes virus infections. *J Antimicrob Chemother* 1983.(12 Suppl. B). 1–8.

115. Larson T, Bryson Y. Fomites and herpes simplex virus: the toilet seat revisited. *Pediatr Res* 1982;**16**(4 Pt 2):244A.

116. Thomas 3rd LE, Sydiskis RJ, DeVore DT, Krywolap GN. Survival of herpes simplex virus and other selected microorganisms on patient charts: potential source of infection. *J Am Dent Assoc* 1985;**111**(3):461–4.

117. Douglas Jr RG, Couch RB. A prospective study of chronic herpes simplex virus infection and recurrent herpes labialis in humans. *J Immunol* 1970;**104**(2):289–95.

118. Hatherley LI, Hayes K, Jack I. Herpes virus in an obstetric hospital. II: Asymptomatic virus excretion in staff members. *Med J Aust* 1980;**2**(5):273–5.

119. Finkelhor RS, Lampman JH. Herpes simplex infection following cardiopulmonary resuscitation. *JAMA* 1980;**243**(7):650.

120. Hendricks AA, Shapiro EP. Primary herpes simplex infection following mouth-to-mouth resuscitation. *JAMA* 1980;**243**(3):257–8.

121. Mannis MJ, Wendel RT. Transmission of herpes simplex during cardiopulmonary resuscitation training. *Compr Ther* 1984;**10**(12):15–17.

122. Birt D, Main J. Oral manifestations of herpes simplex virus infections. *Laryngoscope* 1977; **87**(6):872–8.

123. Wald A, Ericsson M, Krantz E, Selke S, Corey L. Oral shedding of herpes simplex virus type 2. *Sex Transm Infect* 2004;**80**(4):272–6.

124. Krone MR, Wald A, Tabet SR, Paradise M, Corey L, Celum CL. Herpes simplex virus type 2 shedding in human immunodeficiency virus-negative men who have sex with men: frequency, patterns, and risk factors. *Clin Infect Dis* 2000;**30**(2):261–7.

125. Kim JS, Nag P, Landay AL, et al. Saliva can mediate HIV-1-specific antibody-dependent cell-mediated cytotoxicity. *FEMS Immunol Med Microbiol* 2006;**48**(2):267–73.

126. Wu F, Zhai W, Ge L, Qi Y, Gao H, Duan K. Incidence of human herpes virus 1-4 type in saliva of 245 human immunodeficiency virus-infected patients. *Hua Xi Kou Qiang Yi Xue Za Zhi* 2012;**30**(5):514–17.

127. Otero RA, Nascimento FN, Souza IP, et al. Lack of association between herpesvirus detection in saliva and gingivitis in HIV-infected children. *Rev Inst Med Trop Sao Paulo* 2015;**57**(3): 221–5.

128. Wang CC, Yepes LC, Danaher RJ, et al. Low prevalence of varicella zoster virus and herpes simplex virus type 2 in saliva from human immunodeficiency virus-infected persons in the era of highly active antiretroviral therapy. *Oral Surg Oral Med Oral Pathol Oral Radiol Endod* 2010;**109**(2):232–7.

129. Andersson-Ellstrom A, Svennerholm B, Forssman L. Prevalence of antibodies to herpes simplex virus types 1 and 2, Epstein–Barr virus and cytomegalovirus in teenage girls. *Scand J Infect Dis* 1995;**27**(4):315–18.

130. Gershon AA, Gershon MD. Pathogenesis and current approaches to control of varicella-zoster virus infections. *Clin Microbiol Rev* 2013;**26**(4):728–43.

131. Leclair JM, Zaia JA, Levin MJ, Congdon RG, Goldmann DA. Airborne transmission of chickenpox in a hospital. *N Engl J Med* 1980;**302**(8):450–3.

132. Levin MJ. Varicella zoster virus and virus DNA in the blood and oropharynx of people with latent or active varicella zoster virus infections. *J Clin Virol* 2014;**61**(4):487–95.

133. Pollak L, Mehta SK, Pierson DL, Sacagiu T, Avneri Kalmanovich S, Cohrs RJ. Varicella-zoster DNA in saliva of patients with meningoencephalitis: a preliminary study. *Acta Neurol Scand* 2015;**131**(6):417–21.

134. Nagel MA, Choe A, Cohrs RJ, et al. Persistence of varicella zoster virus DNA in saliva after herpes zoster. *J Infect Dis* 2011;**204**(6):820–4.

135. Pierson DL, Mehta SK, Gilden D, et al. Varicella zoster virus DNA at inoculation sites and in saliva after zostavax immunization. *J Infect Dis* 2011;**203**(11):1542–5.

136. Birlea M, Cohrs RJ, Bos N, Mehta SK, Pierson DL, Gilden D. Search for varicella zoster virus DNA in saliva of healthy individuals aged 20-59 years. *J Med Virol* 2014;**86**(2):360–2.

137. Stock I. Infectious mononucleosis—a "childhood disease" of great medical concern. *Med Monatsschr Pharm* 2013;**36**(10):364–8.

138. Niederman JC, Miller G, Pearson HA, Pagano JS, Dowaliby JM. Infectious mononucleosis. Epstein–Barr-virus shedding in saliva and the oropharynx. *N Engl J Med* 1976;**294**(25):1355–9.

139. Fafi-Kremer S, Morand P, Germi R, et al. A prospective follow-up of Epstein-Barr virus LMP1 genotypes in saliva and blood during infectious mononucleosis. *J Infect Dis* 2005;**192**(12): 2108–11.

140. Mollbrink A, Falk KI, Linde A, Barkholt L. Monitoring EBV DNA in saliva for early diagnosis of EBV reactivation in solid tumour patients after allogeneic haematopoietic SCT. *Bone Marrow Transplant* 2009;**44**(4):259–61.

141. de Franca TR, de Albuquerque Tavares Carvalho A, Gomes VB, Gueiros LA, Porter SR, Leao JC. Salivary shedding of Epstein–Barr virus and cytomegalovirus in people infected or not by human immunodeficiency virus 1. *Clin Oral Investig* 2012;**16**(2):659–64.

142. Scully C, Porter SR, Di Alberti L, Jalal M, Maitland N. Detection of Epstein–Barr virus in oral scrapes in HIV infection, in hairy leukoplakia, and in healthy non-HIV-infected people. *J Oral Pathol Med* 1998;**27**(10):480–2.

143. Morgan DG, Niederman JC, Miller G, Smith HW, Dowaliby JM. Site of Epstein–Barr virus replication in the oropharynx. *Lancet* 1979;**2**(8153):1154–7.

144. McSherry JA. Myths about infectious mononucleosis. *Can Med Assoc J* 1983;**128**(6):645–6.

145. Henle G, Henle W, Diehl V. Relation of Burkitt's tumor-associated herpes-type virus to infectious mononucleosis. *Proc Natl Acad Sci USA* 1968;**59**(1):94–101.

146. Hoagland RJ. The transmission of infectious mononucleosis. *Am J Med Sci* 1955;**229**(3): 262–72.

147. Crawford DH, Macsween KF, Higgins CD, et al. A cohort study among university students: identification of risk factors for Epstein–Barr virus seroconversion and infectious mononucleosis. *Clin Infect Dis* 2006;**43**(3):276–82.

148. Fidouh-Houhou N, Duval X, Bissuel F, et al. Salivary cytomegalovirus (CMV) shedding, glycoprotein B genotype distribution, and CMV disease in human immunodeficiency virus-seropositive patients. *Clin Infect Dis* 2001;**33**(8):1406–11.

149. Carstens J, Andersen HK, Spencer E, Madsen M. Cytomegalovirus infection in renal transplant recipients. *Transpl Infect Dis* 2006;**8**(4):203–12.

150. Spector SA, Hirata KK, Newman TR. Identification of multiple cytomegalovirus strains in homosexual men with acquired immunodeficiency syndrome. *J Infect Dis* 1984;**150**(6):953–6.

151. Shepp DH, Match ME, Ashraf AB, Lipson SM, Millan C, Pergolizzi R. Cytomegalovirus glycoprotein B groups associated with retinitis in AIDS. *J Infect Dis* 1996;**174**(1):184–7.

152. Fries BC, Chou S, Boeckh M, Torok-Storb B. Frequency distribution of cytomegalovirus envelope glycoprotein genotypes in bone marrow transplant recipients. *J Infect Dis* 1994; **169**(4):769–74.

153. Grosjean J, Trapes L, Hantz S, et al. Human cytomegalovirus quantification in toddlers saliva from day care centers and emergency unit: a feasibility study. *J Clin Virol* 2014;**61**(3):371–7.

154. Stagno S, Britt B. Cytomegalovirus. In: Remington J, Klein J, Wilson C, Baker C, editors. *Infectious diseases of the fetus and newborn infant*. 6th ed. Philadelphia, PA: Elsevier Saunders; 2006. p. 740–81.

155. Stowell JD, Mask K, Amin M, et al. Cross-sectional study of cytomegalovirus shedding and immunological markers among seropositive children and their mothers. *BMC Infect Dis* 2014;**14**:568.

156. Braun DK, Dominguez G, Pellett PE. Human herpesvirus 6. *Clin Microbiol Rev* 1997;**10**(3): 521–67.

157. Jarrett RF, Clark DA, Josephs SF, Onions DE. Detection of human herpesvirus-6 DNA in peripheral blood and saliva. *J Med Virol* 1990;**32**(1):73–6.

158. Morissette G, Flamand L. Herpesviruses and chromosomal integration. *J Virol* 2010;**84**(23): 12100–9.

159. Tweedy J, Spyrou MA, Pearson M, Lassner D, Kuhl U, Gompels UA. Complete genome sequence of germline chromosomally integrated human herpesvirus 6A and analyses integration sites define a new human endogenous virus with potential to reactivate as an emerging infection. *Viruses* 2016;**8**(1):19.

160. Di Luca D, Mirandola P, Ravaioli T, et al. Human herpesviruses 6 and 7 in salivary glands and shedding in saliva of healthy and human immunodeficiency virus positive individuals. *J Med Virol* 1995;**45**(4):462–8.

161. Cone RW, Huang ML, Ashley R, Corey L. Human herpesvirus 6 DNA in peripheral blood cells and saliva from immunocompetent individuals. *J Clin Microbiol* 1993;**31**(5):1262–7.

162. Wyatt LS, Frenkel N. Human herpesvirus 7 is a constitutive inhabitant of adult human saliva. *J Virol* 1992;**66**(5):3206–9.

163. Hidaka Y, Liu Y, Yamamoto M, et al. Frequent isolation of human herpesvirus 7 from saliva samples. *J Med Virol* 1993;**40**(4):343–6.

164. Ihira M, Yoshikawa T, Ohashi M, et al. Variation of human herpesvirus 7 shedding in saliva. *J Infect Dis* 2003;**188**(9):1352–4.

165. Yoshikawa T, Ihira M, Taguchi H, Yoshida S, Asano Y. Analysis of shedding of 3 beta-herpesviruses in saliva from patients with connective tissue diseases. *J Infect Dis* 2005;**192**(9): 1530–6.

166. Leao JC, Porter S, Scully C. Human herpesvirus 8 and oral health care: an update. *Oral Surg Oral Med Oral Pathol Oral Radiol Endod* 2000;**90**(6):694–704.

167. Leao JC, Caterino-De-Araujo A, Porter SR, Scully C. Human herpesvirus 8 (HHV-8) and the etiopathogenesis of Kaposi's sarcoma. *Rev Hosp Clin Fac Med Sao Paulo* 2002;**57**(4): 175–86.

168. Phipps W, Saracino M, Selke S, et al. Oral HHV-8 replication among women in Mombasa, Kenya. *J Med Virol* 2014;**86**(10):1759–65.

169. Corey L, Brodie S, Huang ML, Koelle DM, Wald A. HHV-8 infection: a model for reactivation and transmission. *Rev Med Virol* 2002;**12**(1):47–63.

170. Pica F, Volpi A. Transmission of human herpesvirus 8: an update. *Curr Opin Infect Dis* 2007;**20**(2):152–6.

171. Triantos D, Horefti E, Paximadi E, et al. Presence of human herpes virus-8 in saliva and non-lesional oral mucosa in HIV-infected and oncologic immunocompromised patients. *Oral Microbiol Immunol* 2004;**19**(3):201–4.

172. Minhas V, Wood C. Epidemiology and transmission of Kaposi's sarcoma-associated herpesvirus. *Viruses* 2014;**6**(11):4178–94.

173. Crabtree KL, Wojcicki JM, Minhas V, et al. Risk factors for early childhood infection of human herpesvirus-8 in Zambian children: the role of early childhood feeding practices. *Cancer Epidemiol Biomarkers Prev* 2014;**23**(2):300–8.

174. Mancuso R, Brambilla L, Boneschi V, et al. Continuous exposure to Kaposi sarcoma-associated herpesvirus (KSHV) in healthcare workers does not result in KSHV infection. *J Hosp Infect* 2013;**85**(1):66–8.

175. Pauk J, Huang ML, Brodie SJ, et al. Mucosal shedding of human herpesvirus 8 in men. *N Engl J Med* 2000;**343**(19):1369–77.

176. Courtois R, Mullet E, Malvy D. Approach to sexuality in an AIDS context in Congo. *Sante* 2001;**11**(1):43–8.

177. Schuster MA, Beckett MK, Corona R, Zhou AJ. Hugs and kisses: HIV-infected parents' fears about contagion and the effects on parent-child interaction in a nationally representative sample. *Arch Pediatr Adolesc Med* 2005;**159**(2):173–9.

178. Smith JW. HIV transmitted by kissing. *Br Med J (Clin Res Ed)* 1987;**294**(6578):1033.

179. Petricciani JC. The biologic possibility of HIV transmission during passionate kissing. *JAMA* 1989;**262**(16):2231.

180. Ho DD, Byington RE, Schooley RT, Flynn T, Rota TR, Hirsch MS. Infrequency of isolation of HTLV-III virus from saliva in AIDS. *N Engl J Med* 1985;**313**(25):1606.

181. Piazza M, Chirianni A, Picciotto L, Cataldo PT, Borgia G, Orlando R. Blood in saliva of patients with acquired immunodeficiency syndrome: possible implication in sexual transmission of the disease. *J Med Virol* 1994;**42**(1):38–41.

182. Woolley R. The biologic possibility of HIV transmission during passionate kissing. *JAMA* 1989;**262**(16):2230.

183. Piazza M, Chirianni A, Picciotto L, et al. Blood in saliva of HIV seropositive drug abusers: possible implication in AIDS transmission. *Boll Soc Ital Biol Sper* 1991;**67**(12):1047–52.

184. Yeung SC, Kazazi F, Randle CG, et al. Patients infected with human immunodeficiency virus type 1 have low levels of virus in saliva even in the presence of periodontal disease. *J Infect Dis* 1993;**167**(4):803–9.

185. Rogers MF, White CR, Sanders R, et al. Lack of transmission of human immunodeficiency virus from infected children to their household contacts. *Pediatrics* 1990;**85**(2):210–14.

186. Friedland GH, Saltzman BR, Rogers MF, et al. Lack of transmission of HTLV-III/LAV infection to household contacts of patients with AIDS or AIDS-related complex with oral candidiasis. *N Engl J Med* 1986;**314**(6):344–9.

187. Centers for Disease Control and Prevention (CDC). Transmission of HIV possibly associated with exposure of mucous membrane to contaminated blood. *MMWR Morb Mortal Wkly Rep* 1997;**46**(27):620–3.

188. Centers for Disease Control and Prevention (CDC). HIV transmission. <https://www.cdc. gov/hiv/basics/transmission.html#ui-id-1>; updated 2016 [accessed 10.01.17].

189. Anonymous. Kissing reported as possible cause of HIV transmission. *J Can Dent Assoc* 1997; **63**(8):603.

190. Fox PC, Wolff A, Yeh CK, Atkinson JC, Baum BJ. Saliva inhibits HIV-1 infectivity. *J Am Dent Assoc* 1988;**116**(6):635–7.

191. Gooch B, Marianos D, Ciesielski C, et al. Lack of evidence for patient-to-patient transmission of HIV in a dental practice. *J Am Dent Assoc* 1993;**124**(1):38–44.

192. Scully C. Viruses in the aetiology of cancer. *Br Dent J* 1988;**164**(11):362–4.

193. Scully C. Viruses and oral squamous carcinoma. *Eur J Cancer B Oral Oncol* 1992;**28B**(1): 57–9.

194. Scully C, Ward-Booth P. Oral carcinoma: evidence for viral oncogenesis. *Br J Oral Maxillofac Surg* 1984;**22**(5):367–71.

195. Campisi G, Panzarella V, Giuliani M, et al. Human papillomavirus: its identity and controversial role in oral oncogenesis, premalignant and malignant lesions (review). *Int J Oncol* 2007;**30**(4):813–23.

196. Maitland NJ, Cox MF, Lynas C, Prime SS, Meanwell CA, Scully C. Detection of human papillomavirus DNA in biopsies of human oral tissue. *Br J Cancer* 1987;**56**(3):245−50.

197. Maitland NJ, Bromidge T, Cox MF, Crane IJ, Prime SS, Scully C. Detection of human papillomavirus genes in human oral tissue biopsies and cultures by polymerase chain reaction. *Br J Cancer* 1989;**59**(5):698−703.

198. Kreimer AR, Villa A, Nyitray AG, et al. The epidemiology of oral HPV infection among a multinational sample of healthy men. *Cancer Epidemiol Biomarkers Prev* 2011;**20**(1):172−82.

199. Hormia M, Willberg J, Ruokonen H, Syrjanen S. Marginal periodontium as a potential reservoir of human papillomavirus in oral mucosa. *J Periodontol* 2005;**76**(3):358−63.

200. Cubie HA. Diseases associated with human papillomavirus infection. *Virology* 2013;**445**(1-2): 21−34.

201. Falter Ii KJ, Frimer M, Lavy D, Samuelson R, Shahabi S. Human papillomavirusassociated cancers as acquired immunodeficiency syndrome defining illnesses. *Rare Tumors* 2013;**5**(2):93−4.

202. Gaester K, Fonseca LA, Luiz O, et al. Human papillomavirus infection in oral fluids of HIV-1-positive men: prevalence and risk factors. *Sci Rep* 2014;**4**:6592.

203. Reusser NM, Downing C, Guidry J, Tyring SK. HPV carcinomas in immunocompromised patients. *J Clin Med* 2015;**4**(2):260−81.

204. Speicher DJ, Ramirez-Amador V, Dittmer DP, Webster-Cyriaque J, Goodman MT, Moscicki AB. Viral infections associated with oral cancers and diseases in the context of HIV: a workshop report. *Oral Dis* 2016;**22**(Suppl. 1):181−92.

205. Del Mistro A, Chieco Bianchi L. HPV-related neoplasias in HIV-infected individuals. *Eur J Cancer* 2001;**37**(10):1227−35.

206. Scully C. Oral squamous cell carcinoma; from an hypothesis about a virus, to concern about possible sexual transmission. *Oral Oncol* 2002;**38**(3):227−34.

207. Scully C. Oral cancer; the evidence for sexual transmission. *Br Dent J* 2005;**199**(4):203−7.

208. D'Souza G, Kreimer AR, Viscidi R, et al. Case-control study of human papillomavirus and oropharyngeal cancer. *N Engl J Med* 2007;**356**(19):1944−56.

209. Herrero R, Castellsague X, Pawlita M, et al. Human papillomavirus and oral cancer: the international agency for research on cancer multicenter study. *J Natl Cancer Inst* 2003;**95**(23): 1772−83.

210. Frisch M, Biggar RJ. Aetiological parallel between tonsillar and anogenital squamous-cell carcinomas. *Lancet* 1999;**354**(9188):1442−3.

211. Hemminki K, Dong C, Frisch M. Tonsillar and other upper aerodigestive tract cancers among cervical cancer patients and their husbands. *Eur J Cancer Prev* 2000;**9**(6):433−7.

212. Hocking JS, Stein A, Conway EL, et al. Head and neck cancer in Australia between 1982 and 2005 show increasing incidence of potentially HPV-associated oropharyngeal cancers. *Br J Cancer* 2011;**104**(5):886−91.

213. Smith EM, Swarnavel S, Ritchie JM, Wang D, Haugen TH, Turek LP. Prevalence of human papillomavirus in the oral cavity/oropharynx in a large population of children and adolescents. *Pediatr Infect Dis J* 2007;**26**(9):836−40.

214. Rintala M, Grenman S, Puranen M, Syrjanen S. Natural history of oral papillomavirus infections in spouses: a prospective Finnish HPV family study. *J Clin Virol* 2006;**35**(1):89−94.

215. Tsao AS, Papadimitrakopoulou V, Lin H, et al. Concordance of oral HPV prevalence between patients with oropharyngeal cancer and their partners. *Infect Agent Cancer* 2016;**11**:21.

216. D'Souza G, Gross ND, Pai SI, et al. Oral human papillomavirus (HPV) infection in HPV-positive patients with oropharyngeal cancer and their partners. *J Clin Oncol* 2014;**32**(23):2408−15.

217. Saranrittichai K, Sritanyarat W, Ayuwat D. Adolescent sexual health behavior in Thailand: implications for prevention of cervical cancer. *Asian Pac J Cancer Prev* 2006;**7**(4):615−18.

218. D'Souza G, Agrawal Y, Halpern J, Bodison S, Gillison ML. Oral sexual behaviors associated with prevalent oral human papillomavirus infection. *J Infect Dis* 2009;**199**(9):1263–9.

219. Pickard RK, Xiao W, Broutian TR, He X, Gillison ML. The prevalence and incidence of oral human papillomavirus infection among young men and women, aged 18-30 years. *Sex Transm Dis* 2012;**39**(7):559–66.

220. Read TR, Hocking JS, Vodstrcil LA, et al. Oral human papillomavirus in men having sex with men: risk-factors and sampling. *PLoS One* 2012;**7**(11):e49324.

221. Antonsson A, Cornford M, Perry S, Davis M, Dunne MP, Whiteman DC. Prevalence and risk factors for oral HPV infection in young Australians. *PLoS One* 2014;**9**(3):e91761.

222. Centers for Disease Control and Prevention (CDC). *HPV and oropharyngeal cancer—fact sheet.* <http://www.cdc.gov/std/hpv/stdfact-hpvandoropharyngealcancer.htm>; updated 2017 [accessed 14.12.16].

223. Kean JM, Rao S, Wang M, Garcea RL. Seroepidemiology of human polyomaviruses. *PLoS Pathog* 2009;**5**(3):e1000363.

224. Robaina TF, Mendes GS, Benati FJ, et al. Polyomavirus in saliva of HIV-infected children, Brazil. *Emerg Infect Dis* 2013;**19**(1):155–7.

225. Baez CF, Guimaraes MA, Martins RA, et al. Detection of Merkel cell polyomavirus in oral samples of renal transplant recipients without Merkel cell carcinoma. *J Med Virol* 2013;**85**(11):2016–19.

226. Martel-Jantin C, Pedergnana V, Nicol JT, et al. Merkel cell polyomavirus infection occurs during early childhood and is transmitted between siblings. *J Clin Virol* 2013;**58**(1):288–91.

227. Burger-Calderon R, Madden V, Hallett RA, Gingerich AD, Nickeleit V, Webster-Cyriaque J. Replication of oral BK virus in human salivary gland cells. *J Virol* 2014;**88**(1):559–73.

228. Berger JR, Miller CS, Mootoor Y, Avdiushko SA, Kryscio RJ, Zhu H. JC virus detection in bodily fluids: clues to transmission. *Clin Infect Dis* 2006;**43**(1):e9–e12.

229. Krone CL, Wyllie AL, van Beek J, et al. Carriage of *Streptococcus pneumoniae* in aged adults with influenza-like-illness. *PLoS One* 2015;**10**(3):e0119875.

230. Abe S, Ishihara K, Adachi M, Sasaki H, Tanaka K, Okuda K. Professional oral care reduces influenza infection in elderly. *Arch Gerontol Geriatr* 2006;**43**(2):157–64.

231. Hui DS, Zumla A. Emerging respiratory tract viral infections. *Curr Opin Pulm Med* 2015;**21**(3):284–92.

232. Sueki A, Matsuda K, Yamaguchi A, et al. Evaluation of saliva as diagnostic materials for influenza virus infection by PCR-based assays. *Clin Chim Acta* 2016;**453**:71–4.

233. World Health Organization (WHO). *Measles.* <http://www.who.int/mediacentre/factsheets/fs286/en/>; updated 2017 [accessed 14.12.16].

234. Maduma-Butshe A, McCarthy N. The burden and impact of measles among the gypsy-traveller communities, Thames valley, 2006-09. *J Public Health (Oxf)* 2013;**35**(1):27–31.

235. Artimos de Oliveira S, Jin L, Siqueira MM, Cohen BJ. Atypical measles in a patient twice vaccinated against measles: transmission from an unvaccinated household contact. *Vaccine* 2000;**19**(9-10):1093–6.

236. Warrener L, Slibinskas R, Chua KB, et al. A point-of-care test for measles diagnosis: detection of measles-specific IgM antibodies and viral nucleic acid. *Bull World Health Organ* 2011;**89**(9):675–82.

237. von Linstow ML, Eugen-Olsen J, Koch A, Winther TN, Westh H, Hogh B. Excretion patterns of human metapneumovirus and respiratory syncytial virus among young children. *Eur J Med Res* 2006;**11**(8):329–35.

238. Senanayake SN. Mumps: a resurgent disease with protean manifestations. *Med J Aust* 2008;**189**(8):456–9.

239. Wharton M, Cochi SL, Hutcheson RH, Schaffner W. Mumps transmission in hospitals. *Arch Intern Med* 1990;**150**(1):47–9.

240. Centers for Disease Control and Prevention (CDC). Updated recommendations for isolation of persons with mumps. *MMWR Morb Mortal Wkly Rep* 2008;**57**(40):1103–5.

241. Gouma S, Hahne SJ, Gijselaar DB, Koopmans MP, van Binnendijk RS. Severity of mumps disease is related to MMR vaccination status and viral shedding. *Vaccine* 2016;**34**(16):1868–73.

242. Chua KB, Lam SK, Goh KJ, et al. The presence of Nipah virus in respiratory secretions and urine of patients during an outbreak of Nipah virus encephalitis in Malaysia. *J Infect* 2001;**42**(1):40–3.

243. Harcourt BH, Lowe L, Tamin A, et al. Genetic characterization of Nipah virus, Bangladesh, 2004. *Emerg Infect Dis* 2005;**11**(10):1594–7.

244. Gurley ES, Montgomery JM, Hossain MJ, et al. Person-to-person transmission of Nipah virus in a Bangladeshi community. *Emerg Infect Dis* 2007;**13**(7):1031–7.

245. Goodgame R. Norovirus gastroenteritis. *Curr Gastroenterol Rep* 2006;**8**(5):401–8.

246. Centers for Disease Control and Prevention (CDC). *Norovirus. Clinical overview.* <https://www.cdc.gov/norovirus/hcp/clinical-overview.html>; updated 2013 [accessed 15.12.17].

247. Le Pendu J, Nystrom K, Ruvoen-Clouet N. Host-pathogen co-evolution and glycan interactions. *Curr Opin Virol* 2014;**7**:88–94.

248. Carmona-Vicente N, Fernandez-Jimenez M, Vila-Vicent S, Rodriguez-Diaz J, Buesa J. Characterisation of a household norovirus outbreak occurred in Valencia (Spain). *BMC Infect Dis* 2016;**16**:124.

249. Morillo SG, Timenetsky Mdo C. Norovirus: an overview. *Rev Assoc Med Bras* 2011;**57**(4): 453–8.

250. Moore MD, Goulter RM, Jaykus LA. Human norovirus as a foodborne pathogen: challenges and developments. *Annu Rev Food Sci Technol* 2015;**6**:411–33.

251. D'Souza DH, Sair A, Williams K, et al. Persistence of caliciviruses on environmental surfaces and their transfer to food. *Int J Food Microbiol* 2006;**108**(1):84–91.

252. Frazer J. Misery-inducing norovirus can survive for months -perhaps years- in drinking water. *Sci Am* 2012.

253. Christie AS. *Norovirus—emetophobia.* <http://www.emetophobiahelp.org/norovirus.html>; updated 2012 [accessed 14.12.16].

254. Levy R, Weissman A, Blomberg G, Hagay ZJ. Infection by parvovirus B 19 during pregnancy: a review. *Obstet Gynecol Surv* 1997;**52**(4):254–9.

255. Duffy C, Woolley Jr. P, Nolting W. Rabies; a case report with notes on the isolation of the virus from saliva. *J Pediatr* 1947;**31**(4):440–7.

256. Helmick CG, Tauxe RV, Vernon AA. Is there a risk to contacts of patients with rabies? *Rev Infect Dis* 1987;**9**(3):511–18.

257. Global Alliance for Rabies Control. *Exposure, prevention and treatment.* <https://rabiesalliance.org/rabies/what-is-rabies-and-frequently-asked-questions/exposure-prevention-treatment>; updated 2017 [accessed 29.05.17].

258. Raza MW, Blackwell CC, Molyneaux P, et al. Association between secretor status and respiratory viral illness. *BMJ* 1991;**303**(6806):815–18.

259. Hilding DA. Literature review: the common cold. *Ear Nose Throat J* 1994;**73**(9):639–43.

260. D'Alessio DJ, Peterson JA, Dick CR, Dick EC. Transmission of experimental rhinovirus colds in volunteer married couples. *J Infect Dis* 1976;**133**(1):28–36.

261. D'Alessio DJ, Meschievitz CK, Peterson JA, Dick CR, Dick EC. Short-duration exposure and the transmission of rhinoviral colds. *J Infect Dis* 1984;**150**(2):189–94.

262. Douglas Jr RG, Rossen RD, Butler WT, Couch RB. Rhinovirus neutralizing antibody in tears, parotid saliva, nasal secretions and serum. *J Immunol* 1967;**99**(2):297–303.

263. Payne DC, Currier RL, Staat MA, et al. Epidemiologic association between FUT2 secretor status and severe rotavirus gastroenteritis in children in the united states. *JAMA Pediatr* 2015;**169**(11):1040–5.

264. Liu Y, Huang P, Tan M, et al. Rotavirus VP8*: phylogeny, host range, and interaction with histo-blood group antigens. *J Virol* 2012;**86**(18):9899–910.

265. Friedman MG, Entin N, Zedaka R, Dagan R. Subclasses of IgA antibodies in serum and saliva samples of newborns and infants immunized against rotavirus. *Clin Exp Immunol* 1996;**103**(2):206–11.

266. Van Trang N, Vu HT, Le NT, Huang P, Jiang X, Anh DD. Association between norovirus and rotavirus infection and histo-blood group antigen types in Vietnamese children. *J Clin Microbiol* 2014;**52**(5):1366–74.

267. Domonova EA, Shipulina OI, Kuevda DA, et al. Detection of rubella virus RNA in clinical material by real time polymerase chain reaction method. *Zh Mikrobiol Epidemiol Immunobiol* 2012;(1):60–7.

268. Brajao de Oliveira K. Torque teno virus: a ubiquitous virus. *Rev Bras Hematol Hemoter* 2015;**37**(6):357–8.

269. Gardner CL, Ryman KD. Yellow fever: a reemerging threat. *Clin Lab Med* 2010;**30**(1):237–60.

270. Chancey C, Grinev A, Volkova E, Rios M. The global ecology and epidemiology of West Nile Virus. *Biomed Res Int* 2015;**2015**:376230.

271. Kelley RE, Berger JR, Kelley BP. West Nile Virus meningo-encephalitis: possible sexual transmission. *J La State Med Soc* 2016;**168**(1):21–2.

272. Scully C, Samaranayake LP. Emerging and changing viral diseases in the new millennium. *Oral Dis* 2016;**22**(3):171–9.

273. World Health Organization. Zika virus infection: global update on epidemiology and potentially associated clinical manifestations. *Wkly Epidemiol Rec* 2016;**91**(7):73–81.

274. Barzon L, Pacenti M, Berto A, et al. Isolation of infectious Zika virus from saliva and prolonged viral RNA shedding in a traveller returning from the Dominican Republic to Italy, January 2016. *Euro Surveill* 2016;**21**(10):30159.

275. Tauro LB, Bandeira AC, Ribeiro GS, et al. Potential use of saliva samples to diagnose Zika virus infection. *J Med Virol* 2017;**89**(1):1–2.

276. Froeschl G, Huber K, von Sonnenburg F, et al. Long-term kinetics of Zika virus RNA and antibodies in body fluids of a vasectomized traveller returning from martinique: a case report. *BMC Infect Dis* 2017;**17**(1):55.

277. Grischott F, Puhan M, Hatz C, Schlagenhauf P. Non-vector-borne transmission of Zika virus: a systematic review. *Travel Med Infect Dis* 2016;**14**(4):313–30.

278. Foy BD, Kobylinski KC, Chilson Foy JL, et al. Probable non-vector-borne transmission of Zika virus, Colorado, USA. *Emerg Infect Dis* 2011;**17**(5):880–2.

279. Siqueira WL, Moffa EB, Mussi MC, Machado MA. Zika virus infection spread through saliva—a truth or myth? *Braz Oral Res* 2016;**30**(1). e46.

280. Leao JC, Gueiros LA, Lodi G, Robinson NA, Scully C. Zika virus: oral healthcare implications. *Oral Dis* 2017;**23**(1):12–17.

281. D'Ortenzio E, Matheron S, Yazdanpanah Y, et al. Evidence of sexual transmission of Zika virus. *N Engl J Med* 2016;**374**(22):2195–8.

CHAPTER 5

Fungi, Protozoa, Parasites, and Other Infective Agents Transmissible by Kissing

In this chapter we review relevant nonviral and nonbacterial agents; many of these are seen especially in immunocompromised people and can be life-changing or even lethal.

Despite initial infection in early childhood and frequent exposure to fungi within the environment, immunocompetent individuals are generally able to contain fungi or maintain the yeast in a latent state. However, immune deficiencies lead to disseminating infections that, in the case of systemic (deep) mycoses, are often fatal without effective rapid clinical intervention.[1] Infections can be transmitted by various routes but fungal infections such as candidiasis usually are transmitted by close mucosal contacts, though genetic factors play a role.

Those living in developing settings, minority groups, institutionalized persons, and the immunocompromised remain at high risk of protozoan diseases. Although protozoa which reside within the blood or internal organs of the host are not usually transmitted from one person to another, the fecal–oral route may play a relevant role.[2] Even though they are not considered as a major mode of transmission of protozoa, kissing contaminated hands and some sexual practices, such as oral–anal contacts, are susceptible to transmit protozoa.

Prions–infectious particles, which are the causative agents of Transmissible Spongiform Encephalopathies (TSE) such as scrapie, variant Creutzfeldt–Jakob Disease (vCJD), and bovine spongiform encephalopathy, may be introduced via intracerebral, intravenous, intraperitoneal, or intraventricular infection. Moreover, interspecies oral transmission has been demonstrated.[3]

Fungi, protozoa, and prions may be present in saliva in quantities possibly sufficient to infect other individuals.[4] Consequently, the transmission of these infectious agents by kissing is biologically plausible.

Saliva Protection and Transmissible Diseases. DOI: http://dx.doi.org/10.1016/B978-0-12-813681-2.00005-6
© 2017 Elsevier Inc. All rights reserved.

5.1 FUNGI

Fungi are ubiquitous. Many human pathogenic fungi are soil-inhabiting species but, under appropriate conditions, i.e., if the person is immunocompromised or suitable local conditions exist, many can become opportunistic. Most fungi can potentially be harmful, if the immune system is compromised as in diabetes, HIV/AIDS, organ transplants, and cancer treatments.

The most frequent genera in the oral fungal microbiome of healthy people are *Candida* species—up to 75% of the population—followed by *Aspergillus, Aureobasidium, Cladosporium, Cryptococcus, Fusarium,* and *Saccharomycetales.*[5] Candidosis (candidiasis or moniliasis)-caused by any type of *Candida*—a dimorphic fungus (it grows as both mycelium and yeasts), is the main issue in oral healthcare especially where salivation or immunity are defective, but even fungi such as the deep mycoses (e.g., *Mucor*) can infect the orofacial region.

Superficial fungal infections are caused by fungi such as *Tinea* spp. that attack skin or its appendages (nail, feathers, and hair) and/or mucosae. Intermediate fungal infections (e.g., *Candida* spp., *Dermophyton,* and *Tricophyton*) may extend to a considerable depth within the tissues, but, unlike systemic mycoses rarely affect other locations. Systemic mycoses enter the body through an internal organ or a deep focus, such as the lungs, the digestive tract, or the paranasal sinuses. Infections can spread through the blood, causing disseminated disease. There are two types of systemic mycoses: opportunistic infections (systemic candidiasis, aspergillosis, and mucormycosis) and endemic respiratory infections (histoplasmosis, blastomycosis, coccidioidomycosis, paracoccidioidomycosis, and cryptococcosis).[6]

5.2 CANDIDAL COLONIZATION AND CANDIDOSIS

Candida albicans (*C. albicans*) is the major infectious agent of candidiasis, is one of the most common intermediate infections, and is commonly found in the mouth, digestive tract, and vagina of apparently perfectly healthy people, but occasionally, may affect the skin, nails, mouth, bronchii, and lungs. *C. albicans* produces extracellular proteases such as surface-associated aspartic proteases, Sap9 and Sap10, which function in cell surface integrity and cell separation during budding.[7]

Candida spp. oral colonization of infants usually arises from the mother, frequently by kissing the child on the lips.[8] Gastrointestinal tract colonization is also influenced by breast milk bioactive factors. Up to 15% of infants aged until 6 months have fairly stable oral colonization. Infants with a furry pet at home had a lower frequency of *Candida* at 3 months, but virtually all older children are colonized.[9] Intimate or deep kissing and oral sex might transmit *Candida* to/from the genitals or mouth from one person to another.[10] *Candida* is also transmitted sexually; oral intercourse may predispose to recurrent vaginal candidiasis.[11] Sexual behavior, rather than the presence of *Candida* species at various body locations of the male partner, is associated with recurrences of *C. albicans* vulvovaginitis.[12]

Cigarette smoke has effects on oral commensals, especially *Candida*, which can cause oral candidosis.[13] Inhaled corticosteroids can also precipitate oral candidiasis.[14] *Candida* oral colonization and infections are also increased by predisposing factors such as antibiotic use, hyposalivation, oral prostheses, and a low CD4+ T lymphocyte count.[15] Increases in the *Candida* load correlates positively with class *Bacilli*, and negatively with class *Fusobacteria*, *Flavobacteria*, and *Bacteroidia* toward dominance by saccharolytic and acidogenic bacteria— *Streptococci*.[16]

5.2.1 Oral Appliances

The most common form of oral candidal infection, denture-related stomatitis, is particularly found with denture age, continuous denture-wearing,[17] and where saliva counts of *C. albicans* and *Lactobacillus* are increased.[18] Denture replacement and denture hygiene improvement aid resolution.[19] Saliva contains proteinaceous factors that detach *C. albicans* cells and lower salivary detachment activity among older people [20] and candidal adhesion increases [21]—possibly one reason that oral candidiasis is more common among older people.

Most patients with denture-related stomatitis generally have intact immunity with normal levels of salivary cystatin-SN, statherin, kininogen-1, desmocollin-2, carbonic anhydrase-6, peptidyl-prolyl *cis*-transisomerase A-like peptides, cystatin C, and several immunoglobulin fragments.[22] The local Th1/Th2 cytokine dichotomy in saliva is not associated with susceptibility to denture-related stomatitis in immunocompetent persons.[23]

Older individuals with denture-related stomatitis however have higher salivary nitric oxide, TGF-β, IL-6, and CCL3 levels and reduced salivary peroxidases, elastase activities, and percentages of salivary TLR4(+) and

CD16(+) neutrophils.[24] Infections caused by *Candida* yeasts are more common in older individuals and usually involve *C. albicans*, with *C. tropicalis*, *C. glabrata*, *C. parapsilosis*, and *C. guilliermondii* in lesser frequencies.[25] Both nonstimulated and stimulated salivary flow rates show a decrease with aging as do secretory immunoglobulin A, lactoferrin transferrin levels, and the neutrophil generation of superoxide and their *Candida* killing activity.[21]

During orthodontic treatment, *Candida* counts increase both in the saliva and on appliance and tooth surfaces.[26] The flora of dental implants with "peri-implantitis" may contain *Candida* spp. in association with *Streptococcus mutans* and *Lactobacilli* and other microorganisms.[27]

5.2.2 Hyposalivation

Patients with xerostomia have decreased whole salivary flow rates, more oral mucosal symptoms, and higher numbers of *Candida* (CFUs); *C. albicans* was isolated mainly from the tongue mucosa and *C. glabrata* was isolated from the angle of the mouth.[28] Candidiasis is especially seen in people with hyposalivation, commonly caused by medications, disorders such as Sjögren's syndrome or ectodermal dysplasia, or as an adverse effect of radiation therapy for treatment of head and neck cancer—*C. albicans*, *C. tropicalis*, *C. krusei*, *C. glabrata*, and *C. parapsilosis* are common.[29–31]

C. dubliniensis, *C. guilliermondii*, *C. lusitaniae*, and *C. kefyr* may also be seen,[32] along with *Enterobacteriaceae*,[30] particularly *Citrobacter*, *Enterobacter*, *Enterococcus*, *Klebsiella*, *Proteus*, and *Pseudomonas*.

5.2.3 Diabetes

Diabetics may suffer oral candidiasis,[33,34] mainly related to a degree of hyposalivation and immune incompetence, and they have a higher candidal carriage rate significantly associated with glycemic control, type of diabetes, and salivary pH compared to controls. *C. albicans* is the most frequently isolated species.[35–37]

Oral *Candida* is higher in patients with prediabetes than in healthy controls, independently of glycemic status.[38]

5.2.4 HIV/AIDS

Oropharyngeal candidiasis (OPC) caused by the commensal organism, *C. albicans*, is the most common oral infection in HIV disease. Although cell-mediated immunity (CMI) by Th1-type CD4+ T-cells is considered the predominant host defense mechanism against OPC, other systemic or local immune mechanisms are critical when blood CD4+ T-cells are reduced below a protective threshold. For example, the Th cytokine profile in saliva may influence resistance or susceptibility to OPC and salivary flow rate, and stimulated whole saliva from HIV-infected patients has decreased anticandidal activity and the lysozyme secretion rate is reduced.[39]

In OPC lesions, CD8+ T-cells become accumulated at the lamina propria–epithelium interface, suggesting some role for CD8+ T-cells against OPC. However, the absence of CD8+ T-cells close to *Candida* at the outer epithelium indicates that susceptibility to OPC involves a dysfunction in the CD8+ T-cells or in the microenvironment. Further evaluation of the buccal mucosa lesion showed that CD8+ T-cell-associated cytokine and chemokine mRNA is increased compared with buccal mucosa from lesion-negative matched controls. The majority of CD8+ T-cells present possess the $\alpha\beta$T-cell receptor and several homing receptors (i.e., 4β7, 4β1, eβ7).

While several adhesion molecules are similar in persons with and without OPC, E-cadherin is reduced in the tissue of those with OPC. These results support evidence for a role of CD8+ T-cells against OPC, but suggest that a putative dysfunction in mucosal T-cell trafficking may be associated with susceptibility to infection. Similar levels of *Candida*-specific antibodies in persons with and without OPC confirmed a limited role for humoral immunity. Finally, oral epithelial cells inhibit the growth of *Candida in vitro* in a static rather than a cidal manner. Clinically, oral epithelial cell anti-*Candida* activity is reduced in HIV+ persons with OPC, compared with controls. The mechanism of action includes a strict requirement for cell contact by an acid-labile moiety on intact, but not necessarily live, epithelial cells, with no role for soluble factors. Taken together, host defense against OPC involves several levels of activity. The status and efficiency of local host defenses when blood CD4+ T-cells are not available appear to play a role in protection against or susceptibility to OPC.[40]

The profile of parotid salivary cytokines is altered as a result of concurrent HIV infection and an oral opportunistic infection, i.e., candidiasis or oral hairy leukoplakia where there are higher levels of IFN-γ than in HIV-infected individuals with no oral disease and significantly higher levels of IL-2, IL-5, and IFN-γ than saliva of healthy controls.[41]

A higher carriage rate and number of colony forming units of *Candida* spp. has been described in saliva of immunocompromised people such as HIV-infected patients.[42] Gender, stage of HIV infection, risk group of HIV infection, systemic disease, and medication use influence hyposalivation and xerostomia,[43] and salivary secretion is reduced in HIV infection [42] further predisposing to candidiasis.

HIV-infected children are colonized by *Candida* species three times more than controls.[44] Absence of highly active antiretroviral therapy (HAART) significantly increased *Candida* spp. colonization.[45] Antiretroviral therapy is associated with decreases in oral candidiasis.[46] HIV protease inhibitors also have a direct effect on secreted aspartyl proteinase (Sap), a key virulence factor of *C. albicans*.[47] Dentinal carious lesions may also be associated with *Candida* spp. colonization in HIV-infected children.[45,48]

Oral candidal carriage is found in 55%−75% of HIV-positive adults [49,50] and often correlates with a CD4 cell number <200 cells/μL.[51] *C. albicans* is the predominant species, and *C. parapsilosis* the most common non-albicans species,[52] with increasing emergence of *C. intermedia*, *C. norvegensis*, and *Rhodotorula rubra* and very small amounts of *Trichosporon mucoides* and *Kodamaea ohmeri*.[53,54] Candida load decreases gradually in the first year of HAART, resulting in a significant reduction in OPC prevalence but without changing the frequency of oral colonization. This is probably due to a decrease in HIV virus load, restoration of the immune system, and *Candida* adherence inhibition by some antiretrovirals.[55]

5.2.5 Transplants

Transplant recipients have oral carriage of *C. albicans* in most cases, and *C. glabrata* is the second most prevalent species.[56,57] There is, as with HIV disease, a correlation between oral *Candida* colonization and (carious) dental status.[57]

Fungal infections carry a greater risk for systemic infection.[58] About 10% of living donor liver transplant patients may develop invasive fungal infections, involving *Candida* spp., *Pneumocystis jiroveci*, or *Aspergillus* spp., the 5-year survival rate being significantly lower in patients with fungal infections than in those without.[59]

Stem cell transplant patients with pretransplant mouth dryness have a higher incidence of oral colonization with *C. albicans* and other fungi than those with normal salivation.[60,61]

5.2.6 Malignant Disease

Oral lesions such as oral carcinomas may harbor *C. albicans* [62,63] and there is a significant correlation with candidal infection rates and the premalignant status of oral leukoplakias.[64]

Patients with hematological malignancies have a roughly 80% incidence of *Candida* colonization with *C. albicans* [65] and stem cell transplants also increase the risk of oral candidiasis.

5.2.7 Down Syndrome

C. albicans can be isolated from the lips [66] and oral cavity of people with Down syndrome.[67] The morphotyping and proteinase production of *C. albicans* isolates from the oral mucosa of children with Down's syndrome differ from those of the isolates of children without Down's syndrome.[68]

5.3 SYSTEMIC FUNGAL INFECTIONS

Systemic fungal infections usually involve saprotrophic fungi found in soil, contracted by inhalation of spores and primarily cause respiratory disease. However, clinical disease is seen mainly in immunocompromised people, and may disseminate via the bloodstream and can be lethal. Human-to-human transmission is uncommon.

5.3.1 Blastomycosis

Blastomycosis ("North American blastomycosis," "Blastomycetic dermatitis," and "Gilchrist's disease") is a fungal infection that apparently arises from soil spores or mycelia, presenting typically as primary pulmonary involvement and skin lesions, which may remain localized, extend, or disseminate. *Blastomyces dermatitidisis*, a fungus that can infect humans, dogs, and occasionally cats, is detected mainly in North America and *Blastomyces brasiliensis* in South America. Blastomycosis is not spread from person-to-person or from animal-to-person, but a case report of a husband and wife both contracting blastomycosis at almost the same time may initiate discussion about human-to-human transmission.[69]

5.3.2 Chromoblastomycosis

Chromoblastomycosis is caused by fungi (*Chaetothyriales*) in soil or by traumatic implantation of contaminated decaying plant material. The infection occurs most frequently on extremities of outdoor workers and usually cause chronic infection in the skin and subcutaneous tissue. Chromoblastomycosis is not spread from person-to-person or from animal-to-person.

5.3.3 Coccidioidomycosis

Coccidioidomycosis is endemic in the south west of the United States (California, Arizona, New Mexico, and Texas), Mexico, and some areas of Central and South America. In the United States, it is most common in the San Joaquin Valley of California (Valley Fever). *Coccidioides immitis* spore inhalation primarily causes respiratory disease, but may disseminate. It is not spread from person-to-person or from animal-to-person, although some cases of transmission through organ transplantation have been described.[70]

5.3.4 Cryptococcosis

Cryptococcosis, also known as *cryptococcal* disease, is caused by *Cryptococcus neoformans* or *Cryptococcus gattii* in soil initially causing respiratory infection. A variety of innate factors interfere with the establishment of cryptococcal infection. Besides physical barriers, such as the skin and the nasal mucosa, the anticryptococcal activity of human serum and saliva has been described. The complement system and phagocytic effector cells are the major players in the nonspecific host immune response to *Cryptococcus*.[71] The cryptococcal polysaccharide capsule is a prominent virulence factor by inhibiting phagocytosis. Salivary immune systems are often effectively defensive.[71,72] The development of an adaptive immune response is essential to overcoming infection though with facultative intracellular pathogens, there is some controversy about the importance of antibody-mediated effective immune defences.[73]

The dramatic course of *Cryptococcus* infections in immunocompromised individuals shows the importance of an intact immune response to the pathogen. Cryptococcosis cases increased with the onset of the AIDS pandemic, the highest incidence still being found in HIV-stricken sub-Saharan Africa, but other immunocompromised individuals (immunosuppressive therapy, organ transplantation, leukemia, lymphoma or sarcoidosis) in whom the organism may disseminate, and has a propensity to localize in the central nervous system, and is potentially fatal.

C. gattii (previously *C. neoformans* var. gattii serotypes B and C), which infects immunocompetent individuals has been isolated from eucalyptus trees in tropical and subtropical regions and, in the United

States in regions of the Pacific Northwest, particularly Oregon and Washington states.

Although natural human-to-human transmission has never been observed, transmission of *Cryptococcus* spp. by solid organ transplantation has been documented.[74]

5.3.5 Histoplasmosis

Histoplasmosis occurs in humans and dogs mainly, the source being soil where *Histoplasma capsulatum* exists as a saprobe especially in droppings of birds and bats. Found worldwide, the incidence is especially high in men from the Mississippi and Ohio River valleys in the United States. Common underlying diseases included HIV infection, diabetes mellitus, and liver diseases.

Histoplasmosis is not communicated from animal-to-animal or person-to-person, but a case of vertical transmission of disseminated histoplasmosis in a mother−infant pair exposed to anti-TNF therapy has been recently reported.[75]

5.3.6 Paracoccidioidomycosis

Paracoccidioidomycosis is a fungal infection endemic to South and Central America, most notably Brazil, Argentina, Colombia, and Venezuela. Paracoccidioidomycosis is caused by the thermally dimorphic fungi *Paracoccidioides brasiliensis* and *Paracoccidioides lutzii*. Although infection is usually subclinical, the fungus can also cause chronic and severe disease,[76] including orofacial lesions.[77−79]

A high level of secretory IgA (sIgA) in saliva of paracoccidioidomycosis patients compared to that of normal donors has been observed, indicating a protective role in neutralizing antigens on mucosal surfaces.[76]

In 1979, a case of a subclinical paracoccidioidomycosis infection in the wife of a patient with paracoccidioidomycosis was reported, whose authors suggested that interhuman contagion appears to be the most feasible source of infection.[80] There are some reports in the literature describing cases of paracoccidioidomycosis infection in organ transplant recipients.[81] However, it seems that paracoccidioidomycosis is not naturally spread from person-to-person or from animal-to-person.

5.3.7 Pneumocystosis

Pneumocystosis, *Pneumocystis carinii* pneumonia or PCP, is caused by *P. jiroveci (carinii)* a fungus ubiquitous in the respiratory tract of mammals, with 90% of adult humans possessing antibodies to the microorganism. *P. jirovecii* is common in saliva of immunocompetent children,[82,83] and is seen especially in HIV disease and cystic fibrosis,[84] hemodialysis patients,[85] and in solid organ transplant recipients.[86]

The primary route of *P. jirovecii* transmission has yet to be proven; however, outbreaks of infection suggest either a possible person-to-person transmission or a common environmental source.[87]

5.4 PROTOZOA

Protozoa found in the oral cavity include *Entamoeba gingivalis* and *Trichomonas tenax* [88] are relatively common in patients with periodontal disease.[89] *Naegleria fowleri* and *Acanthamoeba encephalitis* are amoebae that have been detected in the mouth, and can cause serious forms of meningoencephalitis, but transmission via the oral route has not been demonstrated.[88] Some protozoa can also be recovered from saliva, dental, and periodontal samples from children and teenagers with an intact dentition or with restored teeth.[90]

Texts from the 1970s asserted that transmission could occur via saliva,[91] but this has not been corroborated for these protozoa in subsequent studies.

5.4.1 Leishmaniasis

Protozoa belonging to the genus *Leishmania* cause leishmaniasis— a disease transmitted by sand flies.[92] Clinical presentation of leishmaniasis include visceral, mucosal, and cutaneous involvement. *Leishmania donovani*,[93] *Leishmania infantum*,[94] and *Leishmania siamensis*[95] have been detected in human saliva from patients with leishmaniasis. One study revealed that patients with visceral leishmaniasis had viable pathogenic *Leishmania donovani* in their saliva and that intraperitoneal inoculation of contaminated saliva into hamsters caused the premature death of some and disease in others, leading the authors to state "perhaps the most important natural mode of transmission of kala-azar is from person to person."[93] *Leishmania species* in saliva may have implications for transmission,[96] but person-to-person transmission via saliva has yet to be proven.[88]

5.4.2 Malaria

In virtually all cases, malaria is transmitted through the bite of an infected mosquito, of the genus *Anopheles* which can pass the malaria parasite (*Plasmodium*) through its saliva when it feeds on blood from someone who already has malaria. Only female mosquitoes feed on blood, and so only females transmit malaria. Malaria and dengue are the most common mosquito-borne human-to-human transmitted diseases.[97] Because the malaria parasite is found in red blood cells of an infected person, malaria can also be transmitted through blood transfusion, organ transplant, and via the placenta during pregnancy.[98]

Malaria is not contagious and cannot be contracted from physical contact with an infected person. Malaria cannot be sexually transmitted.[99] The malaria parasite is not in found in infected person's saliva or passed on directly from one person to another.

5.4.3 Toxoplasmosis

Toxoplasmosis is caused by a single-celled parasite *Toxoplasma gondii* the best known zoonotic disease which humans can contract from cats, mainly from their feces. *Toxoplasma* infection occurs through several routes [100]:

- Eating undercooked, contaminated meat (especially pork, lamb, and venison) or other foods or drinks contaminated by utensils, cutting boards, and other foods that have had contact with raw, contaminated meat, or unwashed fruits or vegetables.
- Rarely after receiving a toxoplasma-infected organ transplant or infected blood.
- Mother-to-child (congenital) transmission.

Toxoplasma is found throughout the world, but few infected people have symptoms. However, *toxoplasma* can damage the fetus brain, eyes, or other organs,[101] so pregnant women and immunocompromised people may have serious consequences.

Toxoplasma trophozoites and especially oocysts, the sexual expression of the parasite, facilitate transmission of *Toxoplasma*, as they are much more resistant than the trophozoite.[102,103] *T. gondii* has been found in saliva of some [104] but not all patients with toxoplasmosis.[105]

The role of saliva in transmission of toxoplasmosis is not certain [100,106] but in experimental animals it can be transmitted via saliva.[107]

5.5 PRIONS

Prion diseases, or the TSEs such as Creutzfeldt-Jakob Disease (CJD), are characterized by the conversion of a normal host-coded cellular prion protein into an abnormal protease-resistant isoform that accumulates in, and damages, the brain.[108,109]

Prions are able to spread from the small intestine to salivary glands of sheep.[110] Salivary prion shedding has been detected in sheep and cervids as early as 3 months after ovine scrapie and Chronic Wasting Disease (CWD) exposure and sustained shedding thousands of prion infectious doses throughout the disease course.[111] This provides an additional mechanism for horizontal prion transmission and could have implications for the oral transmission of scrapie and CWD [112] and, presumably, for other prion diseases such as cervix chronic wasting disease and human vCJD.

These findings led to the suggestion that "if this is true for humans, a kiss of a prion may sometimes have lethal consequences".[3] Contemporary *in vitro* assays with the ability to scan for minute quantities of prions have detected prions in the saliva of vCJD-infected humans both during the clinical disease and preclinically.[113] Nevertheless, even using the most sensitive bioassays, there is no hard evidence to date of the transfer of vCJD via saliva.[114]

5.6 CLOSING REMARKS AND PERSPECTIVES

Some nonviral and nonbacterial agents—mainly fungi and protozoa—are seen especially in immunocompromised people and can be life changing or eventually lethal. Most oral colonization of *Candida* spp. in children usually arises from their mothers, frequently by kissing. Most *Candida* spp. isolates fall to nonspecific host immune mediators, and candidiasis results from immune system dysfunction or as a result of local or systemic medical treatment. To date, the most common systemic fungal infections do not seem to spread from person to person with the exception of pneumocystosis. Protozoa have been isolated in saliva but person-to-person transmission via saliva has yet to be proven. Prion diseases, or transmissible spongiform encephalopathies, are progressive and fatal. Prions have been detected in the saliva of variant Creutzfeldt−Jakob disease−infected humans even preclinically, this finding makes kissing a potential route for transmission of the disease. The search for new biomarkers as tools for diagnosis that are able to detect all transmissible prion diseases even at preclinical stages of infection is desirable but not yet possible, and remains a formidable challenge.

REFERENCES

1. Pilmis B, Puel A, Lortholary O, Lanternier F. New clinical phenotypes of fungal infections in special hosts. *Clin Microbiol Infect* 2016;**22**(8):681–7.

2. de Graaf M, Beck R, Caccio SM, et al. Sustained fecal-oral human-to-human transmission following a zoonotic event. *Curr Opin Virol* 2017;**22**:1–6.

3. Da Costa Dias B, Weiss SF. A kiss of a prion: new implications for oral transmissibility. *J Infect Dis* 2010;**201**(11):1615–16.

4. Kang JG, Kim SH, Ahn TY. Bacterial diversity in the human saliva from different ages. *J Microbiol.* 2006;**44**(5):572–6.

5. Ghannoum MA, Jurevic RJ, Mukherjee PK, et al. Characterization of the oral fungal microbiome (mycobiome) in healthy individuals. *PLoS Pathog.* 2010;**6**(1):e1000713.

6. Carrasco-Zuber JE, Navarrete-Dechent C, Bonifaz A, Fich F, Vial-Letelier V, Berroeta-Mauriziano D. Cutaneous involvement in the deep mycoses: a review. Part II—Systemic mycoses. *Actas Dermosifiliogr* 2016;**107**(10):816–22.

7. Albrecht A, Felk A, Pichova I, et al. Glycosylphosphatidylinositol-anchored proteases of *Candida albicans* target proteins necessary for both cellular processes and host-pathogen interactions. *J Biol Chem* 2006;**281**(2):688–94.

8. de SA, Janaína V, Hahn RC. Prevalence of *Candida* spp in the oral cavity of infants receiving artificial feeding and breastfeeding and the breasts of nursing mothers. *J Pediatric Infect Dis* 2011;**6**(4):231–6.

9. Stecksen-Blicks C, Granstrom E, Silfverdal SA, West CE. Prevalence of oral candida in the first year of life. *Mycoses.* 2015;**58**(9):550–6.

10. Mansur AT, Aydingoz IE, Artunkal S. Facial *Candida folliculitis*: possible role of sexual contact. *Mycoses.* 2012;**55**(2):e20–2.

11. Hellberg D, Zdolsek B, Nilsson S, Mardh PA. Sexual behavior of women with repeated episodes of vulvovaginal candidiasis. *Eur J Epidemiol* 1995;**11**(5):575–9.

12. Reed BD, Zazove P, Pierson CL, Gorenflo DW, Horrocks J. *Candida* transmission and sexual behaviors as risks for a repeat episode of *Candida vulvovaginitis*. *J Womens Health (Larchmt)* 2003;**12**(10):979–89.

13. Soysa NS, Ellepola AN. The impact of cigarette/tobacco smoking on oral candidosis: an overview. *Oral Dis.* 2005;**11**(5):268–73.

14. Fukushima C, Matsuse H, Saeki S, et al. Salivary IgA and oral candidiasis in asthmatic patients treated with inhaled corticosteroid. *J Asthma.* 2005;**42**(7):601–4.

15. Menezes Rde P, Borges AS, Araujo LB, Pedroso Rdos S, Roder DV. Related factors for colonization by *Candida* species in the oral cavity of HIV-infected individuals. *Rev Inst Med Trop Sao Paulo* 2015;**57**(5):413–19.

16. Kraneveld EA, Buijs MJ, Bonder MJ, et al. The relation between oral candida load and bacterial microbiome profiles in dutch older adults. *PLoS One.* 2012;**7**(8):e42770.

17. Bilhan H, Sulun T, Erkose G, et al. The role of *Candida albicans* hyphae and *Lactobacillus* in denture-related stomatitis. *Clin Oral Investig* 2009;**13**(4):363–8.

18. Abaci O, Haliki-Uztan A, Ozturk B, Toksavul S, Ulusoy M, Boyacioglu H. Determining *Candida* spp. incidence in denture wearers. *Mycopathologia.* 2010;**169**(5):365–72.

19. Pires FR, Santos EB, Bonan PR, De Almeida OP, Lopes MA. Denture stomatitis and salivary candida in Brazilian edentulous patients. *J Oral Rehabil* 2002;**29**(11):1115–19.

20. Kamagata-Kiyoura Y, Abe S, Yamaguchi H, Nitta T. Reduced activity of *Candida* detachment factors in the saliva of the elderly. *J Infect Chemother* 2004;**10**(1):59–61.

21. Tanida T, Ueta E, Tobiume A, Hamada T, Rao F, Osaki T. Influence of aging on candidal growth and adhesion regulatory agents in saliva. *J Oral Pathol Med* 2001;**30**(6):328−35.

22. Bencharit S, Altarawneh SK, Baxter SS, et al. Elucidating role of salivary proteins in denture stomatitis using a proteomic approach. *Mol Biosyst.* 2012;**8**(12):3216−23.

23. Leigh JE, Steele C, Wormley F, Fidel Jr PL. Salivary cytokine profiles in the immunocompetent individual with *Candida*-associated denture stomatitis. *Oral Microbiol Immunol* 2002;**17**(5): 311−14.

24. Gasparoto TH, Sipert CR, de Oliveira CE, et al. Salivary immunity in elderly individuals presented with candida-related denture stomatitis. *Gerodontology.* 2012;**29**(2):e331−9.

25. de Resende MA, de Sousa LV, de Oliveira RC, Koga-Ito CY, Lyon JP. Prevalence and antifungal susceptibility of yeasts obtained from the oral cavity of elderly individuals. *Mycopathologia.* 2006;**162**(1):39−44.

26. Arslan SG, Akpolat N, Kama JD, Ozer T, Hamamci O. One-year follow-up of the effect of fixed orthodontic treatment on colonization by oral candida. *J Oral Pathol Med* 2008;**37**(1): 26−9.

27. Alcoforado GA, Rams TE, Feik D, Slots J. Microbial aspects of failing osseointegrated dental implants in humans. *J Parodontol.* 1991;**10**(1):11−18.

28. Shinozaki S, Moriyama M, Hayashida JN, et al. Close association between oral *Candida* species and oral mucosal disorders in patients with xerostomia. *Oral Dis* 2012;**18**(7):667−72.

29. Jham BC, Franca EC, Oliveira RR, Santos VR, Kowalski LP, da Silva, et al. Candida oral colonization and infection in Brazilian patients undergoing head and neck radiotherapy: a pilot study. *Oral Surg Oral Med Oral Pathol Oral Radiol Endod* 2007;**103**(3):355−8.

30. Gaetti-Jardim EJ, Ciesielski FI, de Sousa FR, Nwaokorie F, Schweitzer CM, Avila-Campos MJ. Occurrence of yeasts, pseudomonads and enteric bacteria in the oral cavity of patients undergoing head and neck radiotherapy. *Braz J Microbiol* 2011;**42**(3):1047−55.

31. Zhang YY, Li AQ, Wang NN, Liu LN, Cui JL. Oral Candida species distribution in patients receiving radiotherapy for head and neck cancer. *Shanghai Kou Qiang Yi Xue* 2014;**23**(5): 605−8.

32. de Freitas EM, Nobre SA, Pires MB, Faria RV, Batista AU, Bonan PR. Oral candida species in head and neck cancer patients treated by radiotherapy. *Auris Nasus Larynx* 2013;**40**(4):400−4.

33. Kadir T, Pisiriciler R, Akyüz S, Yarat A, Emekli N, Ipbüker A. Mycological and cytological examination of oral candidal carriage in diabetic patients and non-diabetic control subjects: thorough analysis of local aetiologic and systemic factors. *J Oral Rehabil* 2002;**29**(5):452−7.

34. Soysa NS, Samaranayake LP, Ellepola AN. Diabetes mellitus as a contributory factor in oral candidosis. *Diabet Med* 2006;**23**(5):455−9.

35. Al-Attas SA, Amro SO. Candidal colonization, strain diversity, and antifungal susceptibility among adult diabetic patients. *Ann Saudi Med* 2010;**30**(2):101−8.

36. Huang JH, Liu Y, Liu HW. Comparative study on oral candidal infection in individuals with diabetes mellitus and impaired glucose regulation. *Zhonghua Kou Qiang Yi Xue Za Zhi* 2012; **47**(6):335−9.

37. Balan P, Gogineni SB, Kumari NS, et al. Candida carriage rate and growth characteristics of saliva in diabetes mellitus patients: a case-control study. *J Dent Res Dent Clin Dent Prospects* 2015;**9**(4):274−9.

38. Javed F, Ahmed HB, Mehmood A, Saeed A, Al-Hezaimi K, Samaranayake LP. Association between glycemic status and oral candida carriage in patients with prediabetes. *Oral Surg Oral Med Oral Pathol Oral Radiol* 2014;**117**(1):53−8.

39. Lin AL, Johnson DA, Patterson TF, et al. Salivary anticandidal activity and saliva composition in an HIV-infected cohort. *Oral Microbiol Immunol* 2001;**16**(5):270−8.

40. Fidel Jr. PL. Candida-host interactions in HIV disease: relationships in oropharyngeal candidiasis. *Adv Dent Res* 2006;**19**(1):80–4.

41. Black KP, Merrill KW, Jackson S, Katz J. Cytokine profiles in parotid saliva from HIV-1-infected individuals: changes associated with opportunistic infections in the oral cavity. *Oral Microbiol Immunol* 2000;**15**(2):74–81.

42. Jainkittivong A, Lin AL, Johnson DA, Langlais RP, Yeh CK. Salivary secretion, mucin concentrations and *Candida* carriage in HIV-infected patients. *Oral Dis* 2009;**15**(3):229–34.

43. Nittayananta W, Chanowanna N, Jealae S, Nauntofte B, Stoltze K. Hyposalivation, xerostomia and oral health status of HIV-infected subjects in thailand before HAART era. *J Oral Pathol Med* 2010;**39**(1):28–34.

44. Alves TP, Simoes AC, Soares RM, Moreno DS, Portela MB, Castro GF. Salivary lactoferrin in HIV-infected children: correlation with *Candida albicans* carriage, oral manifestations, HIV infection and its antifungal activity. *Arch Oral Biol* 2014;**59**(8):775–82.

45. Cerqueira DF, Portela MB, Pomarico L, de Araujo Soares RM, de Souza IP, Castro GF. Oral *Candida* colonization and its relation with predisposing factors in HIV-infected children and their uninfected siblings in Brazil: the era of highly active antiretroviral therapy. *J Oral Pathol Med* 2010;**39**(2):188–94.

46. Pomarico L, Cerqueira DF, de Araujo Soares RM, et al. Associations among the use of highly active antiretroviral therapy, oral candidiasis, oral candida species and salivary immunoglobulin A in HIV-infected children. *Oral Surg Oral Med Oral Pathol Oral Radiol Endod* 2009;**108**(2):203–10.

47. Tsang CS, Hong I. HIV protease inhibitors differentially inhibit adhesion of *Candida albicans* to acrylic surfaces. *Mycoses.* 2010;**53**(6):488–94.

48. Cerqueira DF, Portela MB, Pomarico L, Soares RM, de Souza IP, Castro GF. Examining dentinal carious lesions as a predisposing factor for the oral prevalence of candida spp in HIV-infected children. *J Dent Child (Chic)* 2007;**74**(2):98–103.

49. Castro G, Martinez R. Relationship between serum and saliva antibodies to candida and isolation of candida species from the mucosa of HIV-infected individuals. *Mycoses.* 2009;**52**(3):246–50.

50. Kamtane S, Subramaniam A, Suvarna P. A comparative study of oral candidal carriage and its association with CD4 count between HIV-positive and healthy individuals. *J Int Assoc Provid AIDS Care* 2013;**12**(1):39–43.

51. Slavinsky III J, Myers T, Swoboda RK, Leigh JE, Hager S, Fidel Jr. PL. Th1/Th2 cytokine profiles in saliva of HIV-positive smokers with oropharyngeal candidiasis. *Oral Microbiol Immunol* 2002;**17**(1):38–43.

52. Jiang L, Yong X, Li R, et al. Dynamic analysis of oral *Candida* carriage, distribution, and antifungal susceptibility in HIV-infected patients during the first year of highly active antiretroviral therapy in guangxi, china. *J Oral Pathol Med* 2014;**43**(9):696–703.

53. Melo NR, Taguchi H, Jorge J, et al. Oral candida flora from Brazilian human immunodeficiency virus-infected patients in the highly active antiretroviral therapy era. *Mem Inst Oswaldo Cruz* 2004;**99**(4):425–31.

54. Junqueira JC, Vilela SFG, Rossoni RD, Barbosa Júnia O, Costa Anna Carolina BP, Rasteiro Vanessa MC, et al. Colonização oral por leveduras em pacientes HIV-positivos no brasil. *Rev Inst Med trop S Paulo* 2012;**54**(1):17–24.

55. Diz Dios P, Ocampo A, Otero I, Iglesias I, Martinez C. Changes in oropharyngeal colonization and infection by *Candida albicans* in human immunodeficiency virus-infected patients. *J Infect Dis* 2001;**183**(2):355–6.

56. Yu J, Zhang M, Wang JL, et al. Candida species distribution in the patients with high risk of deep fungal infections and relevant risk factors: a prospective cohort study. *Zhonghua Yi Xue Za Zhi* 2007;**87**(42):2991–3.

57. Siahi-Benlarbi R, Nies SM, Sziegoleit A, Bauer J, Schranz D, Wetzel WE. Caries-, candida-and candida antigen/antibody frequency in children after heart transplantation and children with congenital heart disease. *Pediatr Transplant* 2010;**14**(6):715–21.

58. Dongari-Bagtzoglou A, Dwivedi P, Ioannidou E, Shaqman M, Hull D, Burleson J. Oral candida infection and colonization in solid organ transplant recipients. *Oral Microbiol Immunol* 2009;**24**(3):249–54.

59. Ohkubo T, Sugawara Y, Takayama T, Kokudo N, Makuuchi M. The risk factors of fungal infection in living-donor liver transplantations. *J Hepatobiliary Pancreat Sci* 2012;**19**(4):382–8.

60. Hermann P, Berek Z, Krivan G, Marton K, Lengyel A. Incidence of oropharyngeal candidosis in stem cell transplant (SCT) patients. *Acta Microbiol Immunol Hung* 2005;**52**(1):85–94.

61. Hermann P, Berek Z, Krivan G, Marton K, Fejerdy P, Lengyel A. Frequency of oral candidiasis in stem cell transplant patients. *Fogorv Sz.* 2006;**99**(1):9–14.

62. Liu K, Gao N, Wang YC, et al. The changes of bacteria group on oral mucosa after radiotherapy of postoperative patients of oral carcinoma. *Hua Xi Kou Qiang Yi Xue Za Zhi* 2005; **23**(2):128–9.

63. Byakodi R, Krishnappa R, Keluskar V, Bagewadi A, Shetti A. The microbial flora associated with oral carcinomas. *Quintessence Int.* 2011;**42**(9):e118–23.

64. Cao J, Liu HW, Jin JQ. The effect of oral candida to development of oral leukoplakia into cancer. *Zhonghua Yu Fang Yi Xue Za Zhi* 2007;**41**(Suppl:):90–3.

65. Lai YY, Bao L, Lu XJ, et al. Candida colonization and invasive fungal infection in hospitalized patients with hematological malignancies. *Zhonghua Yi Xue Za Zhi* 2009;**89**(4):239–42.

66. Scully C, van Bruggen W, Diz Dios P, Casal B, Porter S, Davison MF. Down syndrome: lip lesions (angular stomatitis and fissures) and *Candida albicans*. *Br J Dermatol* 2002;**147**(1): 37–40.

67. Vieira JD, Ribeiro EL, Campos Cde C, et al. *Candida albicans* isolated from buccal cavity of children with Down's syndrome: occurrence and growth inhibition by *Streptomyces* sp. *Rev Soc Bras Med Trop* 2005;**38**(5):383–6.

68. Ribeiro EL, Scroferneker ML, Cavalhaes MS, et al. Phenotypic aspects of oral strains of *Candida albicans* in children with Down's syndrome. *Braz J Biol* 2006;**66**(3):939–44.

69. Bachir J, Fitch GL. Northern wisconsin married couple infected with blastomycosis. *WMJ.* 2006;**105**(6):55–7.

70. Kusne S, Taranto S, Covington S, et al. Coccidioidomycosis transmission through organ transplantation: a report of the OPTN ad hoc disease transmission advisory committee. *Am J Transplant* 2016.

71. Voelz K, May RC. Cryptococcal interactions with the host immune system. *Eukaryot Cell.* 2010;**9**(6):835–46.

72. Igel HJ, Bolande RP. Humoral defense mechanisms in cryptococcosis: substances in normal human serum, saliva, and cerebrospinal fluid affecting the growth of *Cryptococcus neoformans*. *J Infect Dis* 1966;**116**(1):75–83.

73. Harrison TS. *Cryptococcus neoformans* and cryptococcosis. *J Infect.* 2000;**41**(1):12–17.

74. Baddley JW, Schain DC, Gupte AA, et al. Transmission of *Cryptococcus neoformans* by organ transplantation. *Clin Infect Dis* 2011;**52**(4):e94–8.

75. Carlucci JG, Halasa N, Creech CB, et al. Vertical transmission of histoplasmosis associated with anti-tumor necrosis factor therapy. *J Pediatric Infect Dis Soc* 2016;**5**(2):e9–e12.

76. Miura CS, Estevao D, Lopes JD, Itano EN. Levels of specific antigen (gp43), specific antibodies, and antigen-antibody complexes in saliva and serum of paracoccidioidomycosis patients. *Med Mycol.* 2001;**39**(5):423–8.

77. de Almeida O, Jorge J, Scully C, Bozzo L. Oral manifestations of paracoccidioidomycosis (south American blastomycosis). *Oral Surg Oral Med Oral Pathol* 1991;**72**(4):430−5.

78. Sposto MR, Scully C, de Almeida OP, Jorge J, Graner E, Bozzo L. Oral paracoccidioidomycosis. A study of 36 south american patients. *Oral Surg Oral Med Oral Pathol* 1993;**75**(4): 461−5.

79. Almeida OP, Jacks Jr J, Scully C. Paracoccidioidomycosis of the mouth: an emerging deep mycosis. *Crit Rev Oral Biol Med* 2003;**14**(5):377−83.

80. Conti-Diaz IA, Calegari L, Pena de Pereyra M, Casserone S, Fernandez JJ, Scorza L. Paracoccidioidal infection in the wife of a patient with paracoccidioidomycosis. *Sabouraudia.* 1979;**17**(2):139−44.

81. Batista MV, Sato PK, Pierrotti LC, et al. Recipient of kidney from donor with asymptomatic infection by *Paracoccidioides brasiliensis*. *Med Mycol.* 2012;**50**(2):187−92.

82. Contini C, Villa MP, Romani R, Merolla R, Delia S, Ronchetti R. Detection of *Pneumocystis carinii* among children with chronic respiratory disorders in the absence of HIV infection and immunodeficiency. *J Med Microbiol* 1998;**47**(4):329−33.

83. Wakefield AE. Pneumocystis carinii. *Br Med Bull* 2002;**61**:175−88.

84. Sing A, Geiger AM, Hogardt M, Heesemann J. *Pneumocystis carinii* carriage among cystic fibrosis patients, as detected by nested PCR. *J Clin Microbiol* 2001;**39**(7):2717−18.

85. Fritzsche C, Ghanem H, Koball S, Mueller-Hilke B, Reisinger EC. High *Pneumocystis jirovecii* colonization rate among haemodialysis patients. *Infect Dis (Lond)* 2017;**49**(2):132−6.

86. Wintenberger C, Maubon D, Charpentier E, et al. Grouped cases of pulmonary pneumocystosis after solid organ transplantation: advantages of coordination by an infectious diseases unit for overall management and epidemiological monitoring. *Infect Control Hosp Epidemiol* 2017;**38**(2):179−85.

87. Yiannakis EP, Boswell TC. Systematic review of outbreaks of *Pneumocystis jirovecii* pneumonia: evidence that *P. jirovecii* is a transmissible organism and the implications for healthcare infection control. *J Hosp Infect* 2016;**93**(1):1−8.

88. Bergquist R. Parasitic infections affecting the oral cavity. *Periodontology 2000* 2009;**49**:96−105.

89. Ghabanchi J, Zibaei M, Afkar MD, Sarbazie AH. Prevalence of oral *Entamoeba gingivalis* and *Trichomonas tenax* in patients with periodontal disease and healthy population in Shiraz, Southern Iran. *Indian J Dent Res* 2010;**21**(1):89−91.

90. Vrablic J, Tomova S, Catar G, Randova L, Suttova S. Morphology and diagnosis of *Entamoeba gingivalis* and *Trichomonas tenax* and their occurrence in children and adolescents. *Bratisl Lek Listy* 1991;**92**(5):241−6.

91. Faust EC, Russell PF, Jung RC. *Craig and faust's clinical parasitology*. 8th ed. London, UK: Henry Kimpton; 1970.

92. Corvalan FH, Sampaio RNR, Brustoloni YM, Andreotti R, Lima Júnior MSC. DNA identification of *Leishmania (viannia) braziliensis* in human saliva from a patient with American cutaneous leishmaniasis. *J Venom Anim Toxins incl Trop Dis* 2011;**17**(1):98−102.

93. Forkner CE, Zia LS. Viable *Leishmania donovani* in nasal and oral secretions of patients with kala-azar and the bearing of this finding on the transmission of the disease. *J Exp Med* 1934;**59**(4):491−9.

94. Galai Y, Chabchoub N, Ben-Abid M, et al. Diagnosis of mediterranean visceral leishmaniasis by detection of leishmania antibodies and leishmania DNA in oral fluid samples collected using an oracol device. *J Clin Microbiol* 2011;**49**(9):3150−3.

95. Phumee A, Kraivichian K, Chusri S, et al. Detection of *Leishmania siamensis* DNA in saliva by polymerase chain reaction. *Am J Trop Med Hyg* 2013;**89**(5):899−905.

96. Chusri S, Hortiwakul T, Silpapojakul K, Siriyasatien P. Consecutive cutaneous and visceral leishmaniasis manifestations involving a novel *Leishmania* species in two HIV patients in thailand. *Am J Trop Med Hyg* 2012;**87**(1):76−80.

97. Martina BE, Barzon L, Pijlman GP, et al. Human to human transmission of arthropod-borne pathogens. *Curr Opin Virol* 2017;**22**:13−21.

98. Centers for Disease Control and Prevention (CDC). *Malaria. frequently asked questions.* <https://www.cdc.gov/malaria/about/faqs.html>; updated 2017 [accessed 29.05.17].

99. Malaria. com. *Malaria transmission through sexual contact.* <http://www.malaria.com/questions/malaria-transmission-sex>; updated 2017 [accessed 27.05.17].

100. Tenter AM, Heckeroth AR, Weiss LM. *Toxoplasma gondii*: from animals to humans. *Int J Parasitol* 2000;**30**(12):1217−58.

101. Centers for Disease Control and Prevention (CDC). *Toxoplasmosis. frequently asked questions.* <https://www.cdc.gov/parasites/toxoplasmosis/gen_info/faqs.html>; updated 2013 [accessed 12.01.17].

102. Saari M, Raisanen S. Transmission of acute toxoplasma infection. the survival of trophozoites in human tears, saliva, and urine and in cow's milk. *Acta Ophthalmol (Copenh)* 1974; **52**(6):847−52.

103. Zardi O, Soubotian B. Biology of *Toxoplasma gondii*, its survival in body tissues and liquids, risks for the pregnant woman. *Biochem Exp Biol* 1979;**15**(4):355−60.

104. Amendoeira MR, Coutinho SG. Isolation of *Toxoplasma gondii* from the saliva and tonsils of a three-year-old child. *J Infect Dis* 1982;**145**(4):587.

105. Amato Neto V, Braz LM, Jamra LM, Higaki Y, Pasternak J. Isolation of *Toxoplasma gondii* from the saliva of patients with acquired immunodeficiency syndrome (AIDS). *Rev Hosp Clin Fac Med Sao Paulo* 1990;**45**(4):171−2.

106. Levi GC, Hyakutake S, Neto VA, Correa MO. Presence of *Toxoplasma gondii* in saliva of patients with toxoplasmosis. eventual importance of such verification concerning the transmission of the disease (preliminary report). *Rev Inst Med Trop Sao Paulo* 1968;**10**(1):54−8.

107. Terragna A, Morandi N, Canessa A, Pellegrino C. The occurrence of *Toxoplasma gondii* in saliva. *Tropenmed Parasitol.* 1984;**35**(1):9−10.

108. Smith AJ, Bagg J, Ironside JW, Will RG, Scully C. Prions and the oral cavity. *J Dent Res* 2003;**82**(10):769−75.

109. Scully C, Smith AJ, Bagg J. Prions and the human transmissible spongiform encephalopathies. *Dent Clin North Am* 2003;**47**(3):493−516.

110. Maddison BC, Rees HC, Baker CA, et al. Prions are secreted into the oral cavity in sheep with preclinical scrapie. *J Infect Dis* 2010;**201**(11):1672−6.

111. Henderson DM, Denkers ND, Hoover CE, Garbino N, Mathiason CK, Hoover EA. Longitudinal detection of prion shedding in saliva and urine by chronic wasting disease-infected deer by real-time quaking-induced conversion. *J Virol* 2015;**89**(18):9338−47.

112. Tamguney G, Richt JA, Hamir AN, et al. Salivary prions in sheep and deer. *Prion* 2012;**6**(1): 52−61.

113. Saa P, Cervenakova L. Protein misfolding cyclic amplification (PMCA): Current status and future directions. *Virus Res.* 2015;**207**:47−61.

114. Mathiason CK. Silent prions and covert prion transmission. *PLoS Pathog.* 2015;**11**(12): e1005249.

Printed in the United States
By Bookmasters